Introduction to **Phase Equilibria**

in Ceramics

Clifton G. Bergeron
Subhash H. Risbud
University of Illinois
Urbana, Illinois

The American Ceramic Society
Westerville, Ohio

The American Ceramic Society
P. O. Box 6136
Westerville, Ohio 43086-6136

10 09 08 07 06 05 04   10 9 8 7 6 5 4 3 2 1

ISBN: 1-57498-177-3

Library of Congress Cataloging-in-Publication Data
Bergeron, Clifton G.
        Introduction to phase equilibria in ceramics.
        Bibliography: p.
        1. Ceramics. 2. Phase rule and equilibrium.
        I. Risbud, Subhash H. II. Title
[TP810.B47   1984] 666   84-15748
ISBN: 1-57498-177-3

For information on ordering titles published by the American Ceramic Society, or to request a publications catalog, please call 614-794-5890 or visit www.ceramics.org.

Dedicated to the memory of
ARTHUR L. FRIEDBERG

Preface

It is the intent of the authors to provide a textbook which will serve the needs of an introductory course in phase equilibria designed for students in ceramic engineering and associated disciplines. A quantitative approach is taken throughout the book and the emphasis is placed on the interpretation and prediction of reactions. It has been the experience of the authors that the detailed sample problems are a necessary and important part of teaching the subject at the introductory level. Practice problems are of special benefit and consequently are included at the close of most chapters. The book is also designed to permit self-study for persons who are approaching phase equilibrium calculations for the first time, or are seeking a review.

The growing interest in computer calculations of phase diagrams and the increased awareness of the potential of nonoxide ceramics has prompted us to include brief sections on these topics. We hope that by touching on these areas we will motivate the reader to pursue the extensive and rapidly developing literature in this field, only some of which is listed in the Bibliography for this book.

A number of our colleagues at the University of Illinois contributed to this book over many years. A. I. Andrews first introduced the senior author to the subject, and A. L. Friedberg provided many fruitful discussions, shared his experiences, and made constructive comments. The manuscript in its initial form was used over a period of 10 years as the syllabus for a course in phase equilibria taught to junior-level ceramic engineering students at the University of Illinois. The experience thus gained and the helpful comments of those students guided the authors in putting together the present version.

Urbana, Illinois
February 1984

Clifton G. Bergeron
Subhash H. Risbud

INTRODUCTION TO PHASE EQUILIBRIA IN CERAMICS
Table of Contents

Introduction

An understanding of phase equilibria in ceramic systems is central to the utilization and development of materials in refractories, glass, and other high-temperature technologies. Phase equilibria address significant questions related to the flexibility and constraints, dictated by forces of nature, on the evolution of phase assemblages in ceramics. Phase boundaries also assist in the evaluation of the service stability of a ceramic material, both in the long and short time frames. Thus, knowledge of the stability of a ceramic or glass component in high-temperature or high-pressure environments can often be obtained from an appropriate stable or metastable phase diagram.

In the processing and manufacture of ceramic products, the reactions which occur are understood more clearly if the phase relations under equilibrium conditions are known. The chemical and physical properties of ceramic products are related to the number, composition, and distribution of the phases present. Temperature, pressure, and concentration are the principal variables which determine the kinds and amounts of the phases present under equilibrium conditions. To ceramists, who must understand the effects of these variables on both the processing and the properties of the finished product, the phase equilibrium relations (usually presented in the form of phase diagrams) provide the necessary fundamental information.

The study of phase relations is based on the assumption that the system under consideration is at equilibrium. In the development of reliable information on phase relations, this condition must be satisfied. In a practical sense, however, as in the manufacture or service of a ceramic product, circumstances may not permit a condition of equilibrium to be established. In many cases, it is known that the system is driving toward or approaching equilibrium, and knowledge of the direction in which the reaction is progressing or the direction by which it deviates from equilibrium can be of great value. In some instances involving ceramic processing, the approach to equilibrium actually may be quite close. The progress of a ceramic system toward its stable equilibrium state can often be halted for kinetic reasons, resulting in a phase assembly which can persist metastably for an extended period. The arrest of the equilibrium phases, either inadvertently or by deliberate processing, has given rise to some useful new materials in recent years. Thus, a study of stable and metastable phase equilibrium relations is particularly relevant to ceramic and glass compositions.

While most phase equilibrium diagrams have been and continue to be determined by experimental laboratory techniques, there is a growing trend toward calculation of multicomponent equilibria from thermodynamic data. The validity of many classic ceramic phase equilibrium diagrams, while not basically in doubt, continues to be questioned and revised because the experimental techniques and interpretation of data can vary from one study to another. Nevertheless, the student needs to be aware of both experimental and theoretical methods of determining phase diagrams. The principles of thermodynamics are at the core of much important phase equilibria information. It is thus appropriate for us to begin with a brief review of the definitions and principles of thermodynamics that pertain to phase relations.

The phase rule developed by J. Willard Gibbs was derived from the first and second laws of thermodynamics. If sufficient thermodynamic data were available, equilibrium relations of the phases could be calculated. Usually, such data have not been generated, but nevertheless, the understanding of the thermodynamic basis for the phase rule and the manner by which phase relations can be represented in temperature–composition–pressure diagrams is extremely helpful to ceramists.

1.1. Systems, Phases, and Components

Before proceeding with development of the concept of equilibrium between phases and a derivation of the phase rule, it will be useful to define certain terms commonly used in a treatment of phase equilibria.

System: Any portion of the material universe which can be isolated completely and arbitrarily from the rest for consideration of the changes which may occur within it under varying conditions. Often a system may be considered as composed of smaller systems, which together make up the larger system. For example, consider the reactions between SiO_2 and Al_2O_3. These two materials constitute a system which is called the system Al_2O_3-SiO_2. We could deal with smaller systems within this system, such as the system SiO_2 or the system Al_2O_3, or even a small compositional range, and analyze its behavior with respect to variations in temperature, pressure, and composition.

Phase: Any portion including the whole of a system which is physically homogeneous within itself and bounded by a surface so that it is mechanically separable from any other portions. A separable portion need not form a continuous body as, for example, one liquid dispersed in another. A system may contain one phase or many phases. Phases are distinguished by their different physical characters. The physical character in the physical states of matter—gases, liquids, and crystals—is different. These states of matter, then, are physically distinct and represent different phases. Although gases are completely soluble (miscible) in one another and thus represent only one phase, liquids or crystals may be found to exist in several phases, each phase physically distinct from the others. Water and mercury, for example, are both liquids, each representing a different phase. Crystalline silica, SiO_2, may exist in several crystalline configurations, each consisting of a different phase. As exemplified by solutions, either liquid or crystalline, the homogeneous character of the phase is not confined to a rigid chemical composition, since the existence of a variation in chemical composition of liquid or crystalline phases often does not alter the homogeneous character nor cause any abrupt or distinctive change in the physical structure of the phase. As phases are considered to be distinguished mainly by their physical character, such physical characteristics as density, X-ray diffraction behavior, or optical properties are used to aid in the identification and distinction of phases in a system.

A system composed of only one phase is described as a homogeneous system. A system composed of two or more phases is termed a heterogeneous system.

Components: The components of a system are the smallest number of independently variable chemical constituents necessary and sufficient to express the composition of each phase present in any state of equilibrium. In the alumina-silica system, Al_2O_3 and SiO_2 are the components of the system, since all phases and reactions can be described by using only these two materials. Al, Si, and O would not be the components for they are not the *least* number of chemical substances by which the system can be quantitatively expressed. Consider the reaction:

$$MgCO_3 \rightleftharpoons MgO + CO_2$$

At equilibrium, all three chemical constituents are present, but they are not all components because they are not all independently variable. Choosing any two of the three phases fixes the composition of the remaining phase.

In selecting the components of this system, it is helpful to arrange the information in tabular form (as shown in Table 1.1) where the composition of each phase is expressed in terms of positive, negative, or zero quantities of the chemical constituents chosen as components.

Table 1.1. Chemical Composition of System $MgCO_3 \rightleftharpoons MgO + CO_2$

Phase	Composition of phase in terms of chemical constituents		
	$MgCO_3$ and MgO	MgO and CO_2	$MgCO_3$ and CO_2
$MgCO_3$	$MgCO_3 + 0$ MgO	$MgO + CO_2$	$MgCO_3 + 0$ CO_2
MgO	0 $MgCO_3 +$ MgO	$MgO + 0$ CO_2	$MgCO_3 - CO_2$
CO_2	$MgCO_3 -$ MgO	0 $MgO + CO_2$	0 $MgCO_3 + CO_2$

Table 1.1 shows that any two of the chemical constituents could be selected as the components of the system because each selection resulted in a suitable expression for the composition of each phase present at equilibrium. Note, however, that only one of the selections MgO and CO_2 resulted in an expression involving positive quantities only. In selecting the components of a system, it is preferable to use only positive quantities.

Variance (or degrees of freedom): The number of intensive variables, such as temperature, pressure, and concentration of the components, which must be arbitrarily fixed in order that the condition of the system may be perfectly defined. A system is described as invariant, monovariant, bivariant, etc. according to whether it possesses, respectively, zero, one, two, etc. degrees of freedom.

1.2. Equilibrium

Equilibrium in a system represents a condition in which: (1) the properties of a system do not change with the passage of time, and (2) the same state can be obtained by approaching this condition in more than one manner with respect to the variables of the system.

From a practical point of view, this is an adequate definition. It is somewhat lacking in precision, however, because we have set no limits to the term "time." A small change in properties which is discernible only after many years would indicate that the system is still approaching equilibrium. The approach may be so slow as to be beyond our capability to readily detect a change in some property of the system and, thus, we are led to assume that the system is at equilibrium.

A more precise definition is the thermodynamic definition which states that a system at equilibrium has a minimum free energy. The free energy of a system is the energy available for doing work and is defined by the following relationship:

$$G = E + PV - TS \tag{1.1}$$

where $G =$ Gibbs free energy, $E =$ the internal energy, $P =$ the pressure on the system, $V =$ the volume of the system, $T =$ the absolute temperature, and $S =$ the entropy of the system.

The internal energy, E, of a system represents the total energy of the system, that is, the total of the kinetic and potential energies of all the atoms or molecules in the system. It is not a quantity which can be measured; we

4

can, however, measure differences in internal energy when a system undergoes a change:

$$\Delta E = E_2 - E_1 = q - w \qquad (1.2)$$

where E_2 and E_1 represent the final and initial states of the system, respectively, q is the heat added to the system, and w is the work done by the system.

If a reaction is carried out in a constant-volume calorimeter so that no pressure–volume work is done by the system, then the amount of heat liberated or absorbed by the reaction is a measure of the change in internal energy, ΔE.

If, however, the same reaction is carried out in a constant-pressure calorimeter, the pressure–volume work must be considered:

$$\Delta E = E_2 - E_1 = q - W = q - P(V_2 - V_1)$$

Rearranging terms,

$$E_2 - E_1 = q - PV_2 + PV_1$$

$$q = (E_2 + PV_2) - (E_1 + PV_1)$$

The term $(E + PV)$ is defined as the heat content or enthalpy, H, and is equal to the heat absorbed at constant pressure. H and E are useful terms because they have characteristic and definite values for any system at any definite state of pressure, temperature, and aggregation.

Absolute values of H and E are not known, and consequently, it is necessary to select some arbitrary zero point for these quantities. By convention, the heat contents of elements in their standard states (liquid, solid, gas) at 25°C and 1 atm of pressure are set equal to zero.

In practice, we deal with changes in heat content, ΔH, or changes in internal energy, ΔE. For any change, such as a chemical reaction, the change in heat content is called the heat of reaction and is determined by the difference in heat contents between the products and the reactants.

$$\Delta H = H_{products} - H_{reactants} \qquad (1.3)$$

If ΔH is positive, heat is absorbed during the reaction and the reaction is called endothermic. If ΔH is negative, heat is evolved during the reaction and the reaction is called exothermic.

To raise the temperature of a material, a certain amount of heat must be added. By definition, the heat required to raise the temperature of a given amount of material 1°C is the heat capacity. The units chosen are usually calories per mole per degree. If the heat capacity has been determined at constant volume, the term c_v is used. The heat capacity increases with temperature; the relationship can be expressed by an equation of the form

$$c_p = a + bT + cT^2 + dT^3 \qquad (1.4)$$

where a, b, c, and d are constants.

If the temperature dependency of the heat capacity is known, the change in heat content which occurs during the heating of a given material can be calculated from the following expression:

$$\Delta H = H_{T_2} - H_{T_1} = \int_{T_1}^{T_2} c_p dT \qquad (1.5)$$

Most ceramic reactions are carried out at atmospheric pressure; consequently, the enthalpy changes are more conveniently used in calculations than are changes in internal energy.

The entropy, S, of a system is most easily defined as a measure of the randomness or disorder of a system. A perfect crystal, free of impurities and defects, would have zero entropy at 0 K. As the temperature of the crystal is increased, its atoms absorb energy by thermal motion, some disordering oc-

Table 1.2. Phase-Composition Possibilities

Components	Phases				
	α	β	γ	δ	θ
C_1	X_1^α	X_1^β	X_1^γ	X_1^δ	X_1^θ
C_2	X_2^α	X_2^β	X_2^γ	X_2^δ	X_2^θ
C_3	X_3^α	X_3^β	X_3^γ	X_3^δ	X_3^θ
C_j	X_j^α	X_j^β	X_j^γ	X_j^δ	X_j^θ

curs, and as a consequence, the entropy of the crystal increases to a value characteristic of that temperature and degree of randomness. The entropy of the crystal at some temperature T is given by

$$dS = \frac{dq_r}{T} \tag{1.6}$$

where $dq_r =$ the heat absorbed reversibly.

Because the energy involved in the entropy change is tied up in the random arrangement and thermal motion of the atoms, it is often referred to as unavailable energy.

All spontaneous processes occur with an increase in the entropy of the system and its surroundings; that is, the system goes to a more probable state.

The Gibbs free energy, $G = H - TS$, is a measure of the available energy. It represents the driving force for a reaction. For a reaction to occur spontaneously, the free energy change must be negative (the free energy of the system decreases). When the free energy change is zero ($\Delta G = \Delta H - T\Delta S = 0$), the system is in a state of equilibrium and no further change will occur.

1.3. The Phase Rule

Originally deduced by Gibbs and brought to general use by Roozeboom, the phase rule serves to define the conditions of equilibrium in terms of a relationship between the number of phases and the components of a system.

$$F = C - P + 2 \tag{1.7}$$

where $F =$ the variance or degrees of freedom, $C =$ the number of components, and $P =$ the number of phases.

Every system possesses a certain number of independent variables to which values must be assigned in order to describe the system. Temperature, pressure, composition of phases, magnetic forces, gravitational fields, etc. are variables which may pertain to a given system. In deriving the phase rule, Gibbs considered the following variables: (1) temperature, (2) pressure, and (3) composition (of phases in terms of the components of the system).

To calculate the number of independent variables which a system at equilibrium possesses, we can make a general statement, as follows:

$$\text{Variance} = \begin{bmatrix} \text{Total No. of intensive} \\ \text{variables which a system} \\ \text{possesses} \end{bmatrix} - \begin{bmatrix} \text{No. of intensive variables} \\ \text{which are fixed if the} \\ \text{system is in equilibrium} \end{bmatrix} \tag{1.8}$$

Consider a system of C components existing in P phases, as shown schematically in Fig. 1.1: To describe the composition of each phase, the amount of each component in each phase must be specified. As shown in Table 1.2, the result is represented by PC composition variables. The temperature and pressure must also be specified; thus, the total number of variables is $PC + 2$.

A convenient way to desribe the composition of a phase is to use mole fractions. If a given phase contains n_1 moles of component 1, n_2 moles of component 2, and n_3 moles of component 3, the mole fraction of component

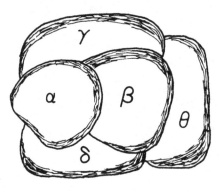

Fig. 1.1. Schematic diagram of system of five phases.

6

1 in the phase is

$$X_1 = \frac{n_1}{n_1 + n_2 + n_3} \qquad (1.9)$$

For each phase the sum of the mole fractions equals unity:

$$X_1 + X_2 + X_3 = 1$$

If all but one mole fraction is specified, the remaining one can be determined by difference. If there are P phases, there are P equations of this type and, therefore, P mole fractions that need not be specified. The total number of variables thus becomes $PC + 2 - P$, which is the first term on the right side of the equation.

To calculate the second term, Gibbs introduced the concept of the thermodynamic or chemical potential which each component possesses. The chemical potential is an intensity term rather than a capacity term. For example, if two phases at different temperatures are brought into contact, heat will flow from the hotter phase to the cooler phase until both are at the same temperature. Each phase may have a different *amount* of heat energy, but thermal equilibrium depends only on the intensity factor, i.e., the temperature.

Similarly, mechanical equilibrium requires that the pressure of all phases be the same; otherwise, one phase will increase in volume at the expense of another until the pressures are equal.

The chemical potential of a component must be the same in all of the phases in which it appears or matter will flow from one phase to another until the chemical potentials are equal. The chemical potential of component i is given by the following expression:

$$\mu_i = \left(\frac{\partial G}{\partial n_i} \right)_{P,T,n_j} \qquad (1.10)$$

where μ = chemical potential, n_i = number of moles of component i, n_j = number of moles of all the other components, G = Gibbs free energy, P = pressure, and T = absolute temperature.

The chemical potential is equivalent to the partial molar free energy of a component in a given phase. In Eq. (1.10) is represented the change in free energy which accompanies the addition or removal of a small increment of component i from a given phase while the temperature, pressure, and the number of moles of all the other components in the phase remain constant. Consider the two-phase system shown in Fig. 1.2. If dn_i moles of component i are transferred from phase β to phase γ while all other variables are held constant, the change in free energy of the β phase during this transfer is given by

$$dG^\beta = \mu_i^\beta dn_i \qquad (1.11)$$

The change in free energy of the γ phase is given by

$$dG^\gamma = -\mu^\gamma dn_i \qquad (1.12)$$

The total change in the free energy of this two-phase system is given by the following expression:

$$dG = dG^\beta + dG^\gamma = (\mu_i^\beta - \mu_i^\gamma)dn_i \qquad (1.13)$$

If the system is at equilibrium, this transfer results in a net free energy change of zero; thus

$$dG = (\mu_i^\beta - \mu_i^\gamma)dn_i = 0$$

Therefore

$$\mu_i^\beta = \mu_i^\gamma$$

and the chemical potential of component i is the same in both phases.

At equilibrium the temperatures of all the phases are equal, their pressures are equal, and the chemical potentials of the respective components are the same between phases. If the concentration of component one in any

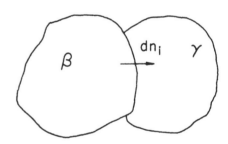

Fig. 1.2. Two-phase system showing dn_i moles of component i transforming from β to γ phase.

phase is specified, the concentration of component one in all of the other phases in which it appears becomes fixed because the chemical potential of the component is the same in each phase. (The concentration of component one is not necessarily the same in each phase, however.) Thus, for each component $(P-1)$, variables are fixed, and since there is a total of C components, the total number of variables which are fixed is $(P-1)C$. Substituting into Eq. (1.8), we derive Eq. (1.14), which is a statement of the phase rule:

$$F = (PC + 2 - P) - (P - 1)C$$
$$F = PC + 2 - P - PC + C$$
$$F = C - P + 2 \qquad\qquad\qquad\qquad\qquad (1.14)$$

Bibliography and Supplementary Reading

K. Denbigh: The Principles of Chemical Equilibrium, 2d ed. Cambridge University Press, England, 1966.

A. Findlay: Phase Rule and Its Applications. Dover, New York, 1951.

J. W. Gibbs, The Scientific Papers of J. Willard Gibbs, Vol. 1, Thermodynamics. Dover, New York, 1961.

H. A. J. Oonk: Phase Theory, The Thermodynamics of Heterogeneous Equilibria. Elsevier, Amsterdam, The Netherlands, 1981.

K. S. Pitzer and L. Brewer: Thermodynamics, 2d ed. Revision of Thermodynamics by G. N. Lewis and M. Randall. McGraw-Hill, New York, 1961.

J. E. Ricci: The Phase Rule and Heterogeneous Equilibrium. Van Nostrand, New York, 1951.

Phase equilibrium relations have been studied for many one-, two-, three-, and multicomponent systems of ceramic materials. For a one-component system, $C = 1$, the statement of the phase rule is as follows:

$$F = 1 - P + 2$$

$$F = 3 - P$$

In a one-component system, a phase experiences two degrees of freedom: $F = 3 - 1 = 2$. The two variables which must be specified to define the system are temperature and pressure. Chemical composition is not a variable, since only one component is being considered. Where two phases ($P = 2$) coexist in a one-component system, there is one degree of freedom, $F = 3 - 2 = 1$. Where three phases ($P = 3$) coexist in a one-component system, the degrees of freedom equal zero, $F = 3 - 3 = 0$, as is the case for point I in Fig. 2.1, which shows a hypothetical one-component system. At point I, three phases (vapor, liquid, and crystals) coexist under equilibrium conditions. Point I is called an invariant point, for there are no degrees of freedom at this point. Point I is also called a triple point, since three phases coexist at this point.

On the boundary lines, IA, IB, and IC, two phases are in equilibrium, and any point on these lines has one degree of freedom. The one degree of freedom on the boundary line indicates that pressure and temperature are dependent on each other and the specification of one of these variables will automatically fix the other variable. Where only one phase exists (an area), pressure and temperature may be varied independently of each other. Thus, in one-component systems, the degree of freedom may be zero, one, or two.

To summarize: three phases coexisting, $F = 3 - 3 = 0$, an invariant point; two phases coexisting, $F = 3 - 2 = 1$, univariant equilibrium (boundary lines); one phase, $F = 3 - 1 = 2$, bivariant equilibrium (areas).

The line AI is called the sublimation curve. It represents the conditions of temperature and pressure under which solid and vapor can coexist. The upper end of the sublimation curve is at the triple point, where solid, liquid, and vapor coexist. The curve IC is called the vaporization curve and represents the temperature–pressure conditions under which liquid and vapor can coexist. In the absence of supercooling, its lower terminus is the triple point. Its upper terminus is the critical point (the temperature above which the gas cannot be liquefied, no matter how great the pressure).

The curve IB is the fusion curve or melting point curve. It represents the line at which solid and liquid are in equilibrium under varied conditions of temperature and pressure. Line IB may be considered as the change in melting point of the solid phase with change in pressure. Increased pressure causes an increase in this melting point. The slope of line IB also indicates the relative density between the crystals and the liquid. A solid is generally more dense than its liquid. This is so in Fig. 2.1, where the liquid–crystal boundary line, IB, slopes with increasing pressure toward the right. Increase in pressure increases density, and from Fig. 2.1 it may be noted that an increase in pressure alone (temperature constant) can cause the liquid to change to its more dense crystalline phase.

2.1. Le Chatelier's Principle

In a qualitative way, one can predict the effect of changes on the equilibrium of a system by means of the principle of Le Chatelier, as follows: "If an attempt is made to change the pressure, temperature, or concentration

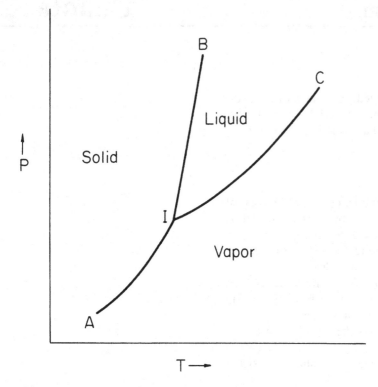

Fig. 2.1. Hypothetical one-component system.

of a system in equilibrium, then the equilibrium will shift in such a manner as to diminish the magnitude of the alteration in the factor which is varied.''

In the example cited previously (solid and liquid in equilibrium), if the volume of the system were held constant and heat were added to the system, the expected result would be an increase in the temperature of the system. According to the Le Chatelier principle, that reaction will occur which tends to diminish the magnitude of the temperature rise. In this case, more of the solid would melt because melting absorbs heat (endothermic). The greater specific volume of the melt would cause an increase in the pressure of the system, and the equilibrium between solid and liquid would thus shift to a position of higher pressure and temperature (toward *B* on curve *IB* of Fig. 2.1).

If the heat content of the solid–liquid system were held constant and the pressure were decreased, a reaction would occur which would tend to diminish the effect of the pressure decrease; that is, the volume would increase. More of the solid would melt, because melting is accompanied by an increase in volume. The heat absorbed by melting causes the temperature of the system to decrease. Thus, the equilibrium between solid and liquid would shift to a position of lower temperature and pressure (toward point *I* on curve *IB* of Fig. 2.1).

Similar examples can be worked out for solid-vapor and for liquid-vapor systems at equilibrium.

A quantitative expression of the principle of Le Chatelier is given by the Clausius–Clapeyron equation which can be related to the slopes of the lines in the pressure-temperature diagram.

$$dP/dT = \Delta H/T\Delta V \qquad\qquad (2.1)$$

where P = pressure on the system, T = the absolute temperature, ΔH = the enthalpy change accompanying the phase change, i.e., solid to liquid, liquid to vapor, etc., usually given in calories per mole, and ΔV = the change in specific volume accompanying the phase change, usually given in cubic centimeters per mole.

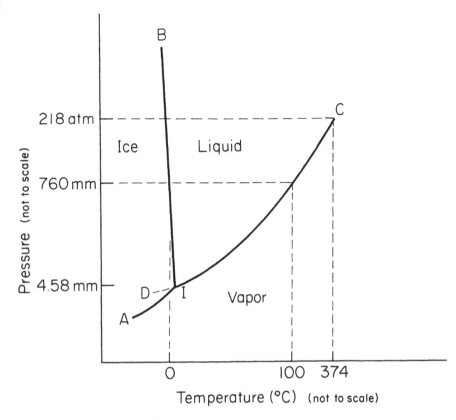

Fig. 2.2. Schematic diagram of part of the water system.

The slopes of the boundary lines are usually positive because, in going from a low to a higher temperature, the enthalpy change is positive and the volume change is most often positive. There are several exceptions involving solid–liquid transformation in which the change in specific volume on heating is not positive and consequently the slopes of the melting point curves are negative. The substances water, bismuth, and antimony exhibit such behavior.

2.2. The Water System

A good example for the discussion of the one-component system is the water system. Such a common material as water is well known, and its different phases, water (the liquid form), ice (the crystalline form), and water vapor (the gaseous form), are very familiar to us at atmospheric pressure. Figure 2.2 shows a schematic diagram of part of the water system. The change at atmospheric pressure on heating ice crystals is noted on the constant pressure (isobaric) line across the diagram. Ice changes to water at 0°C.* Water boils at 100°C.

As pressure on the system is increased, the temperature at which liquid changes to vapor is increased; thus, water boils at a lower temperature at high altitudes where pressure is lower. Decreasing pressure, while it lowers the boiling point, causes the melting point to increase. Generally, however, liquids are less dense than their crystalline form, and the melting points increase with increasing pressure.

The boundary line between liquid and vapor represents the temperature and pressure of boiling points of the liquid or the condensation points of the vapor. As pressure is increased, the boiling point is increased, as is indicated by the line IC in Fig. 2.2. Liquid and vapor are in equilibrium on this line. At point C, called the critical point, the distinction of the two phases is lost; the vapor and liquid have become homogeneous under such high temperatures and pressures. This critical temperature for water is 374.0°C at a critical pressure of 217.7 atm.

*In a closed system and with pure water (i.e., water containing no air in solution), the freezing point of water is actually 0.0099°C at a pressure of 760 mm.

11

Although phase equilibrium diagrams express relationships of materials subjected to the variables temperature, composition, and pressure, it should be recognized that the environment generally experienced by materials is the exposure to the atmosphere and thus to atmospheric pressure. A system in which the materials are exposed to the atmosphere is called an "open" system, whereas a system in which the materials are exposed only to their own vapor pressures is called a "closed" system.

At atmospheric pressure, crystals melt when their vapor pressure (which increases with temperature) equals the vapor pressure of the liquid phase. Some crystals have very high vapor pressures and, at atmospheric pressure, the crystals may evaporate (sublime) completely.

Metastability

The fact that water may be supercooled illustrates a metastable condition. A metastable phase always has a higher vapor pressure than some other phase (which is more stable) in the same temperature range. Supercooled water is a metastable phase, since it has a higher vapor pressure in its temperature range than ice (the more stable phase).

A metastable phase is in a state of metastable equilibrium. One requirement for the stable equilibrium state is that the conditions of equilibrium can be approached in more than one manner. Metastable equilibrium does not comply completely with the definition of equilibrium, since with metastable equilibrium, the method of approach is usually through only one definite procedure. Supercooled water, for example, can be obtained only by cooling water and not by heating ice. Still, supercooled water fulfills the other equilibrium requirement in that the phase experiences no further change with the passage of time.

The addition of a stable phase tends to transform the metastable phase to the stable phase. The addition of ice, the stable phase, will cause the spontaneous transformation of the supercooled water to ice, but only by some external influence such as a disturbance or seeding (i.e., the addition of surface energy) will the metastable phase revert to a more stable phase. Supercooled water as a metastable phase has a certain region of temperature in which it may remain as water; that is, the degree of supercooling is limited to a certain minimum temperature (point D, Fig. 2.2) below which supercooled water reverts to ice crystals.

Instability or the term "unstable phase" is often confused with metastability or metastable phase. Instability appears more as a general term describing such phenomena as slow, sluggish reactions or the decomposition of certain materials. Metastability, however, refers specifically to the special conditions under which metastable phases exist.

2.3. Hypothetical Systems

The construction of phase diagrams from statements of the physical relations between phases has proved to be a valuable aid in gaining familiarity with phase diagrams. A typical exercise is given below:

Problem: A one-component system consists of two solid phases (S_1 and S_2), a liquid phase (L), and a vapor phase (V). S_1 and S_2 can be melted and sublimed. S_1 is more dense than S_2, which is more dense than the liquid. The transformation $S_1 \rightarrow S_2$ is endothermic. Construct the phase diagram which describes the system.

Solution: Because both solids can be sublimed and melted, each solid phase must have a common boundary with the liquid phase and with the vapor phase. Thus, the diagram must contain two distinct sublimation curves and two distinct fusion curves in addition to the vaporization curve and the solid–solid transformation boundary. A possible arrangement is shown in Fig. 2.3.

The relative positions of the phase areas can be determined by the principle of Le Chatelier. For example, consider the changes which occur along the

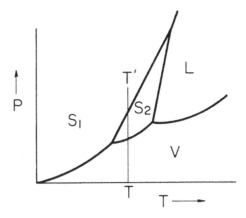

Fig. 2.3. Solution to problem.

isothermal line T-T' in Fig. 2.3 as the pressure is increased. One would expect, as pressure is increased, that the molar volume would decrease, i.e., change to a more dense phase. Thus, the vapor phase condenses to a solid (S_2), and with further increases in pressure, the solid S_2 transforms to S_1, which is the denser of the two phases.

The slopes of the lines can be determined by employing the Clausius-Clapeyron equation, $dP/dT = \Delta H/(T\Delta V)$. Consider the changes which occur on heating the solid phase S_1 while holding the pressure constant. Because the transformation $S_1 \rightarrow S_2$ is endothermic the ΔH term is positive. The change in molar volume, ΔV, is also positive because S_2 is less dense than S_1. The slope of the boundary line between S_1 and S_2 is thus positive, as shown in Fig. 2.3. The same type of analysis can be applied to the other boundary lines in the system.

The volume change which occurs when the liquid is vaporized is given by $V_v - V_\ell$. Because $V_v \gg V_\ell$, ΔV can be set equal to V_v, and if the vapor behaves as an ideal gas

$$V_v = \frac{RT}{P} \tag{2.2}$$

The slope of the vaporization curve is given by

$$\frac{dP}{dT} = \frac{\Delta H_v P}{RT^2} \tag{2.3}$$

$$\int \frac{dP}{P} = \int \frac{\Delta H_v}{RT^2} dT$$

$$\ln P = -(\Delta H_v/RT) \tag{2.4}$$

Similarly, for the sublimation curve

$$P = e^{-(\Delta H_{sub}/RT)} \tag{2.5}$$

The vaporization and sublimation curves each have an exponential temperature dependency and thus slope continuously upward.

For the melting reaction and for solid–solid transformation, the ΔV term does not change greatly with temperature and consequently the fusion curves and the boundaries between solid phases are nearly straight lines. Where three lines come together at a triple point, the metastable extension of any

13

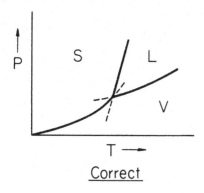
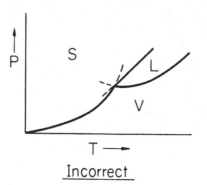

Correct   Incorrect

Fig. 2.4. Construction at triple point.

line must lie between the equilibrium curves of the other two lines (see Fig. 2.4).

Justification for this rule becomes apparent if one compares the slopes of the lines at the triple point. For the transformation solid→liquid, the molar volume change is very small and the slope $dP/dT = \Delta H/(T\Delta V)$ has a relatively large value compared with that for the solid→vapor transformation, where the slope is small owing to the large molar volume change which accompanies vaporization of the liquid. The enthalpy of vaporization is always smaller than the enthalpy of sublimation ($\Delta H_{sub} = \Delta H_{vap} + \Delta H_{fus}$); the ΔV terms for sublimation and vaporization are nearly equal, and consequently the slope of the vaporization curve must be less than that of the sublimation curve at the triple point. Thus, the intersections must occur as shown in the diagram on the left in Fig. 2.4.

2.4. The Silica System

Most ceramic crystals have very high melting points, and the vapor pressures of these crystals are very low and have not been measured. However, the phase relations representing the effect of the temperature on phase changes can be drawn since the relative stability of the phases is known. Such is the diagram of the SiO_2 system shown in Figs. 2.5 and 2.6. These are one-component diagrams in which the temperature ranges of the stable and metastable phases are shown. Vapor pressure plotted on the vertical axis has not been measured, but since the stability of phases is known, the representation with respect to temperature and pressure can be drawn schematically.

Figures 2.5 and 2.6 show the many polymorphic forms of SiO_2 crystals. Polymorphism is the term used to indicate the existence of the same chemical composition in two or more crystalline forms, each having its own characteristic vapor pressures and temperature ranges of stability and metastability and being distinguished by its different crystalline structure and physical properties. The terms polymorphism and allotropism are often used interchangeably to describe the phenomenon of the existence of the same chemical composition in several crystalline forms. Polymorphism is usually employed as a general term for this phenomenon, whereas allotropism refers to this phenomenon in elemental substances, such as sulfur, iron, or tin.

The polymorphic forms of silica are present in several stable and metastable forms. One method of representing these temperature regions of stability and metastability is as shown in Fig. 2.5. The stable form of quartz at room temperature is α-quartz. The α-quartz crystals when heated to 573 °C will change rapidly to β-quartz and, on cooling, the β-quartz will revert to the α-quartz form. This is a reversible change that always occurs. Stable β-quartz when heated to 870 °C may change to β-tridymite. However, this inversion is sluggish, and β-quartz may exist in a metastable form at high temperatures, where it will change to a metastable β-cristobalite or melt to liquid silica. This

14

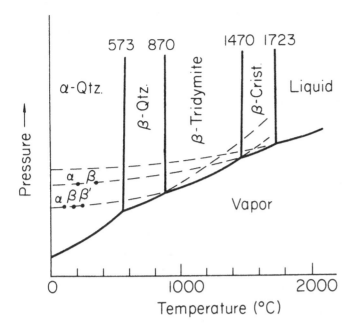

Fig. 2.5. Stability relations in the silica system at atmospheric pressure (open system). Crystalline phases are α- and β-quartz, α-, β-, β'-, and β"-tridymite, and α- and β-cristobalite. Dashed lines indicate metastable phases (after C. N. Fenner, *Am. J. Sci.,* **36** [214] 331–84 (1913)).

molten silica can be supercooled, remaining as silica glass at room temperatures.

Polymorphic changes in crystals are sluggish or rapid, depending on the amount of structural change occurring between the polymorphic forms. Transformations between polymorphic forms which differ markedly in crystalline structure (i.e., complete alteration of the space lattice) are sluggish. On the other hand, inversions from one crystalline form to another that is quite similar in crystalline structure tend to be rapid. Thus, the α- to β-quartz inversions or the α- to β-cristobalite inversions, which are associated with a slight modification of structure,[†] are rapid. The quartz-to-tridymite inversion is sluggish since it entails a major change in crystalline structure. Note also that the rapid inversions are reversible, whereas the sluggish inversions tend toward irreversibility of the two polymorphic forms and the formation of metastable forms.

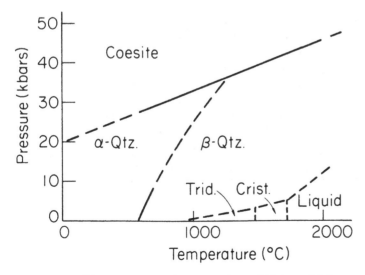

Fig. 2.6. Phase diagram for system SiO_2 at higher pressures (after F. C. Kracek, "Polymorphism"; in Encyclopedia Brittanica, 1953).

[†]A distinct change in geometrical symmetry of the space lattice, but atom arrangement is affected only slightly.

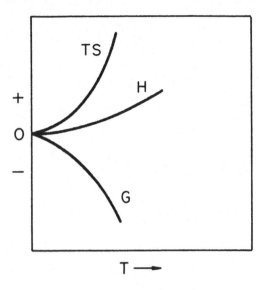

Fig. 2.7. Temperature dependency of free energy of a phase.

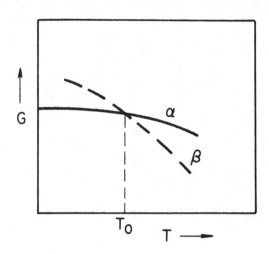

Fig. 2.8. Free energy-temperature relation of polymorphic phases. T_0 = temperature of phase transformation.

The stable phase at a given temperature is the one with the lowest free energy. The free energy at constant pressure and temperature is given by

$$G = H - TS \qquad (2.6)$$

where G = Gibbs free energy, H = enthalpy, S = entropy, and T = absolute temperature.

As the temperature is increased, both the enthalpy and the entropy increase (Fig. 2.7), but the entropy term increases at a greater rate; consequently, the free energy of a substance decreases as the temperature is increased.

At a phase transformation temperature (such as 573 °C for the α-→β-quartz transformation), the free energies of the two phases are equal (Fig. 2.8).

$$(G_\beta - G_\alpha) = \Delta H - T\Delta S = 0$$

$$\Delta H = T\Delta S \qquad (2.7)$$

The heat of transformation is a representation of the change in entropy or increase in the disorder as the phase transforms to its more probable configuration. As the temperature is increased above the transformation temperature, the crystalline phase with the lower free energy is the stable phase. Associated with polymorphic transformation are the changes in physical properties, such as density, crystalline structure, refractive index, birefringence, and color. With the rapid inversion of α- to β-quartz or α- to β-cristobalite, the change in volume is so large that care must be exercised in the firing or service of ceramic bodies containing these crystals to avoid disruption of the body. In the manufacture of silica brick, for example, an important goal is to convert a large percentage of the quartz to tridymite so that this damaging α- to β-quartz reaction will not appear again. Mineralizers and fluxes are often used to effect the formation of some polymorphic form; for example, in silica brick, calcium oxide is added to the quartz materials to promote the formation of higher temperature forms of silica and to promote the vitreous bond in the brick.

2.5. The Titania and Zirconia Systems

Other ceramic materials exhibiting polymorphism include titania (TiO_2) and zirconia (ZrO_2). The stability relations in the titania system are shown schematically in Fig. 2.9. Anatase, the low-temperature tetragonal modifica-

16

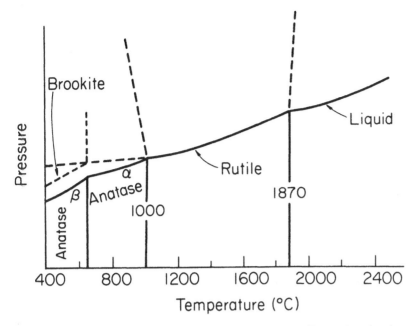

Fig. 2.9. Stability relations in the titania system (from data by A. Schröder, *Z. Kristallogr.*, **66**, 493 (1928); *ibid.*, **67**, 485 (1928); E. N. Bunting, *J. Res. Natl. Bur. Stand.* (*U.S.*), **11** [5] 719–25 (1933)).

tion, exists in two forms, β- and α-anatase, which change reversibly at 642 °C. At approximately 1000 °C, α-anatase inverts irreversibly to rutile, which is also tetragonal but of different lattice parameters. Another crystalline modification of titania is brookite, found principally as a natural material. When brookite is heated, it changes directly to rutile at approximately 650 °C. Rutile melts at 1870 °C.

In the zirconia system, the polymorphic forms of zirconia may be related as shown in Fig. 2.10. The low-temperature modification is a monoclinic crystal corresponding to the mineral baddeleyite. This crystal changes reversibly to a tetragonal modification at approximately 1000 °C. The tetragonal

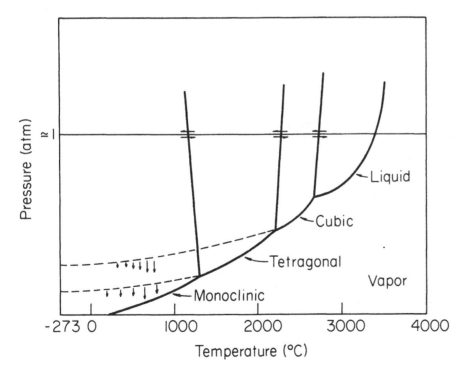

Fig. 2.10. Proposed diagram for system ZrO_2 (after R. Ruh and T. J. Rockett, *J. Am. Ceram. Soc.*, **53** [6] 360 (1970)).

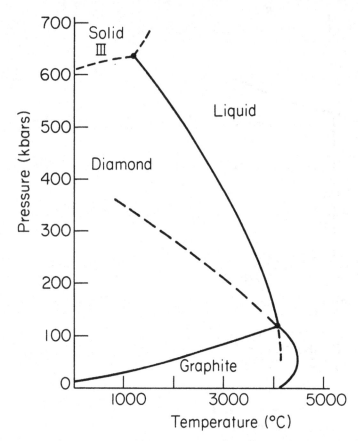

Fig. 2.11. Proposed diagram for carbon system.

modification is stable over 1000 °C and metastable below this temperature. A large volume change (approximately 7%) accompanies the monoclinic–tetragonal inversion. At 2300 °C, tetragonal zirconia changes to cubic zirconia, which is stable to the melting point, 2715 °C. For a practical refractory, small amounts of CaO, Y_2O_3, or MgO are added to the ZrO_2 to promote greater stability of this cubic form over the entire temperature range.

2.6. The Carbon System

There has been much interest in the carbon system because of the knowledge that diamond is one of the denser forms of graphite. Attempts to manufacture diamonds by subjecting carbon to high temperatures and pressures formed the basis for a myriad of experiments which began about the year 1800. The first successful production of diamonds from graphite was announced by the General Electric Co. in 1955 after many years of arduous research effort. The conditions for the formation of diamonds are shown on the carbon phase diagram, Fig. 2.11. For the direct conversion of graphite to diamond, temperatures above 4000 K and pressures approaching 150 kbars are required. By the use of suitable catalysts, the conversion can be made at considerably lower temperatures and pressures.

Most diamonds for industrial purposes are small, usually less than 0.5 mm in diameter. They are used in many grinding, polishing, and sawing operations. The ceramic industries use about one-third of all the industrial diamonds produced. Many of the fusion-cast refractories are trimmed to close tolerances by diamond-impregnated cutting wheels.

Problems for Chapter 2

2.1. *A one-component system consists of the following phases: 1 liquid (L), 1 vapor (V), and 3 solids (S_1, S_2, and S_3). Construct the P–T diagram which satisfies the following conditions:*

    S_1 and S_2 can be melted and sublimed.

    S_3 can be melted but not sublimed.

    S_3 is denser than S_1, S_2, and the liquid.

    The liquid is denser than S_2 which is denser than S_1.

    The transformations $S_1 \to S_2$ and $S_1 \to S_3$ are both endothermic.

2.2. *If a system is found to obey the phase rule, may we assume that the system is at equilibrium? Explain your answer.*

2.3. *Classify the following systems as monovariant, divariant, or invariant. Explain your answers.*
   a. *Alpha quartz in equilibrium with beta quartz at the transition temperature.*
   b. *Monoclinic zirconia at room temperature.*
   c. *Ice in equilibrium with its vapor.*

2.4. *The following phases are known to exist in a system at constant pressure:*

    Cordierite ($2MgO \cdot 2Al_2O_3 \cdot 5SiO_2$)

    Mullite ($3Al_2O_3 \cdot 2SiO_2$)

    Forsterite ($2MgO \cdot SiO_2$)

    Protoenstatite ($MgO \cdot SiO_2$)

    Periclase (MgO)

   a. *What are the components of the system?*
   b. *Could all of the above-listed phases coexist at equilibrium? Explain your answer.*

2.5. *At 970°C the following equation represents an equilibrium reaction:*

$$CaCO_3 \leftrightarrows CaO + CO_2, \quad \Delta H = +39 \ kcal/mol$$

From a consideration of LeChatelier's principle, in which direction would you expect the equilibrium to shift if heat were added to the system? If the pressure on the system were decreased? Explain how you arrived at your answers.

Bibliography and Supplementary Reading

F. R. Boyd and J. L. England, *J. Geophys. Res.*, **65**, 752 (1960).

P. W. Bridgman, *J. Chem. Phys.*, **5**, 965 (1937) (The Water System, *P–T* diagram).

F. P. Bundy, *J. Chem. Phys.*, **38**, 618 (1963).

E. N. Bunting, *J. Res. Natl. Bur. Std.*, **11**, 719 (1933).

F. Dachille and R. Roy, *Z. Kristallogr. Krystallgeom., Kristallphys., Kristallchem.*, **111**, 455 (1959).

C. N. Fenner, *Am. J. Sci.*, **36**, 331 (1913).

D. R. Gaskell: pp. 155–85 in Introduction to Metallurgical Thermodynamics, Scripta, Washington, DC, 1973.

R. B. Gordon, Principles of Phase Diagrams in Materials Systems. McGraw-Hill, New York, 1968.

F. C. Kracek, "Polymorphism"; in Encyclopedia Brittanica, 1953.

R. Ruh and T. J. Rockett, *J. Am. Ceram. Soc.*, **53** [6] 360 (1970).

A. Schroder, *Z. Kristallogr., Kristallgeom., Kristallphys., Kristallchem.*, **66**, 493 (1928).

A. Schroder, *Z. Kristallogr., Kristallgeom., Kristallphys., Kristallchem.*, **67**, 485 (1928).

R. B. Sosman: The Phases of Silica, Rutgers University Press, NJ, 1965.

R. A. Swalin: Thermodynamics of Solids, 2d ed. Wiley, New York, 1972.

Chapter 3

The Two-Component System

A system consisting of only two components is called a *binary system* or *two-component system*. The two-component system is influenced by three variables: temperature, pressure, and composition. Representation of a two-component system with these three variables requires a three-dimensional model such as that shown schematically in Fig. 3.1. The end-members, or components, of the system are A and B and are represented by means of the pressure–temperature diagram of the kind discussed in Chapter 2.

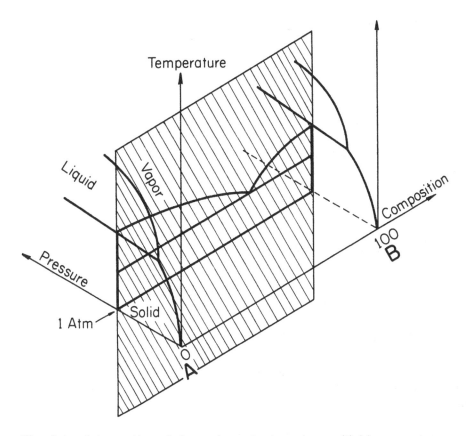

Fig. 3.1.   Intersection of plane of constant pressure with binary system.

In most systems of ceramic materials, the phase relations are determined at atmospheric pressure. In Fig. 3.1 a plane has been passed through the figure at a position corresponding to a constant pressure of 1 atm. The intersection of this plane with the fusion curves of the end-members and with those of all the intervening mixtures of A and B gives the phase diagram for a two-component system at constant pressure. If pressure is omitted as a variable, the number of variables in a system is two: temperature and composition. The phase rule reduces to $F = C - P + 1$ and in this form is referred to as the "condensed phase rule" or the "phase rule for condensed systems."

In a closed system at atmospheric pressure, the vapor phase would not appear until very high temperatures were reached. Diagrams are seldom extended to these temperatures. In an open system, those atoms which acquire sufficient energy to escape the surface of the melt or solid will be able to diffuse to the surroundings and may condense on cooler surfaces. These physical relations are illustrated in Fig. 3.2.

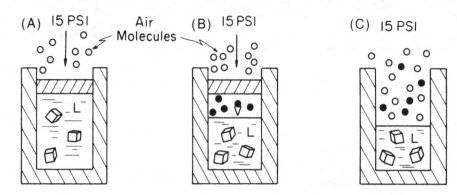

Fig. 3.2. Physical relations in closed and open systems: (A) closed system of solid/liquid at temperature T_1; (B) closed system in which vapor phase appears when $V_P \geq 15$ psi and temperature $T_2 >> T_1$; (C) open system in which some liquid molecules escape from the system.

A phase diagram for a condensed binary system is shown in Fig. 3.3. This kind of diagram is often called a constitution-temperature diagram. The composition of the two components is represented on the abscissa and the temperature on the ordinate. In this hypothetical system all phases that may form between components A and B at various temperatures are represented. The freezing points of 100% A and 100% B are shown. Additions of B to A cause the freezing point to be lowered from M toward E. This curved line ME is called the *liquidus line*. The liquidus temperature is the temperature above which no crystals can exist. It is the locus of temperatures at which crystals first begin to appear on cooling the melt under equilibrium conditions. Below line FG no liquid exists, the composition being entirely crystalline. This line is called the *solidus line*.

The application of the phase rule to this system can be demonstrated as follows:

1. For a single phase, such as liquid, the number of variables which must be specified to define the system can be determined from the phase rule;

$$F = C - P + 1$$

$$F = 2 - 1 + 1 = 2$$

Both the temperature and the composition must be specified to uniquely describe the liquid phase.

2. If both liquid and solid phases are present, such as A + liquid or B + liquid, the number of variables which must be specified to define the system is $F = 2 - 2 + 1 = 1$. Either the composition or the temperature is all that is necessary to define the system in terms of the variables which the system possesses. If the composition of the liquid phase is given (the liquid phase in this system is the only phase of variable composition), the temperature is automatically fixed because there is only one temperature at which that particular liquid composition is in equilibrium with a solid phase. An isothermal line drawn through the two-phase region shown in Fig. 3.3 connects the compositions of the solid and liquid phase which are in equilibrium at that temperature. This line is called a *tie line* or *conode* and connects the *conjugate* phases (equilibrium phases). In Fig. 3.3 the lines A_1X_1, A_2M_2, A_3M_3, and N_1B_1 are all tie lines.

If the temperature is the variable specified, the liquid composition in equilibrium with crystals of A or with crystals of B is automatically fixed. The two-phase region may be thought of as containing an infinite number of tie lines, the extremities of which indicate the compositions of the two phases which are in equilibrium at that particular temperature.

Referring again to Fig. 3.3, consider those compositions which lie in the range 0–62% B. At temperature T_3, any sample within that compositional range will be composed of liquid and crystals of A. The composition of the liquid is given by the intersection of the tie line A_3M_3 with the liquidus line

22

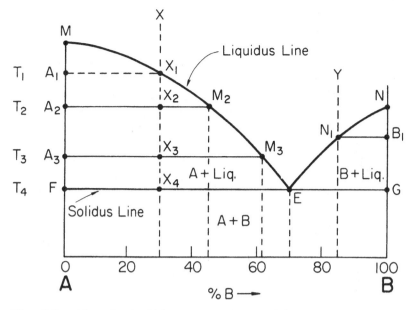

Fig. 3.3.   Hypothetical binary system containing one eutectic.

and is equal to 62% B and 38% A. The relative proportions of liquid and crystals of A will be determined by the gross composition of the sample and can be calculated from a relation called the lever rule, which is derived as follows:

At temperature A' consider a sample of gross or overall composition given by point S on the composition axis of Fig. 3.4.

Let L = the weight fraction of liquid of composition $\begin{cases} X_A^l = \dfrac{RB}{AB} \\[2mm] X_B^l = \dfrac{AR}{AB} \end{cases}$

where X_A^l is the weight fraction of A in the liquid phase and X_B^l that of B

$(1-L)$ = the weight fraction of solid of composition $\begin{cases} X_A^s = \dfrac{AB}{AB} \\[2mm] X_B^s = 0 \end{cases}$

W = the weight of the sample of composition $\begin{cases} X_A = \dfrac{SB}{AB} \\[2mm] X_B = \dfrac{AS}{AB} \end{cases}$

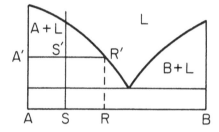

Fig. 3.4.   Binary system A-B illustrating the lever rule.

The total weight of A in the sample is equal to the weight of A in the liquid plus the weight of A in the solid phase.

$$X_A W = X_A^l L W + X_A^s (1-L) W$$

$$X_A = X_A^l L + X_A^s (1-L)$$

$$X_A = X_A^l L + X_A^s - X_A^s L$$

$$X_A^s - X_A = L(X_A^s - X_A^l)$$

$$L = \frac{X_A^s - X_A}{X_A^s - X_A^l} = \frac{AB - SB}{AB - RB}$$

$$(1-L) = 1 - \frac{AB - SB}{AB - RB}$$

$$(1-L) = \frac{SB - RB}{AB - RB}$$

23

$$\frac{L}{(1-L)} = \frac{AB - SB}{SB - RB} = \frac{AS}{SR} = \frac{\text{amount of liquid}}{\text{amount of solid}}$$

The relation may be visualized as a lever in which the fulcrum is at S' and the lever arms are $A'S'$ and $S'R'$ for the liquid and solid, respectively.

Tracing a cooling melt through a two-phase region will help in understanding the significance of the solidus and liquidus lines and the use of the two rules described in the preceding paragraphs. In Fig. 3.3 a melt of composition X is to be cooled. Melt X is completely liquid on cooling down to point X_1. At point X_1, which is on the liquidus curve, crystals of A begin to appear. Equilibrium conditions along the liquidus line involve an infinitesimal quantity of crystals in the melt. Cooling the melt from X_1 to X_2 causes a measurable amount of crystals to form. The amount and composition of the crystals and melt present at this temperature T_2 are determined by the use of a tie line and the lever rule. The tie line A_2M_2 is drawn as shown in the figure. The intersections of this isothermal line with the boundaries of the two-phase area give the composition of the two equilibrium phases. At point X_2, for example, the horizontal line intersects the boundaries of the two-phase area at points A_2 and M_2. The composition at point A_2 consists of crystals of A. The composition of the melt at point M_2 is 55% A and 45% B.

The relative amounts of the two phases present can be determined by use of a lever relation. Point X_2, for example, is the fulcrum of a lever. One lever arm is A_2X_2, the distance from point A_2 to point X_2. The other lever arm is X_2M_2. The relative amount of A crystals is given by the lever arm X_2M_2 divided by the sum of the lever arms, A_2M_2. Thus, the percentage of phases at point X_2 is given by

$$\% \text{ of A crystals} = \frac{X_2M_2}{A_2M_2} \times 100 = \frac{15 \text{ units}}{45 \text{ units}} \times 100 = 33\%$$

$$\% \text{ of melt} = \frac{A_2X_2}{A_2M_2} \times 100 = \frac{30 \text{ units}}{45 \text{ units}} \times 100 = 67\%$$

At point X_3, a similar procedure is used to determine the composition and percentage of phases. The composition of the melt is represented by point M_3. The amounts of the two phases, A crystals and melt, are given by

$$\% \text{ of A crystals} = \frac{X_3M_3}{A_3M_3} \times 100 = \frac{32.5}{62.} \times 100 = 52\%$$

$$\% \text{ of melt} = \frac{A_3X_3}{A_3M_3} \times 100 = \frac{30}{62.5} \times 100 = 48\%$$

composition of melt $M_3 = 37.5\%$ A and 62.5% B

These calculations on a cooling melt may be listed as given in Table 3.1.

The liquidus lines (ME and NE in Fig. 3.5) can also represent solubility curves. Consider the isothermal line RF at temperature T_1. If a small amount of B is added to crystals of A, a small quantity of a liquid of composition K is formed. With further additions of B, more liquid of composition K is formed and the amount of crystalline A is decreased correspondingly. When the gross composition of the sample corresponds to the point K, only an infinitesimal quantity of crystalline A remains in equilibrium with the melt. Continued additions of B to the mixture cause the liquid composition to change from K toward G, at which point the liquid becomes saturated in B and crystals of B begin to precipitate from the melt. From point G to point F the liquid composition remains constant and is given by point G; the sample contains an increasing proportion of solid B as the gross composition of the sample approaches that of point F. Thus, the liquidus line ME may be thought of as the solubility limit for A in the melt, and the line NE may be thought of as the solubility limit for B in the melt.

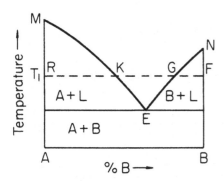

Fig. 3.5. Liquidus lines also represent solubility limits.

Table 3.1.   Calculations on Cooling Melt X, 70% A, 30% B

Temp (°C)	Proportions (phases)	% (Phases)	% Composition of phases (in terms of components)	Analysis A	B
T_1	30 units melt	100	$\begin{cases} A = 70 \\ B = 30 \end{cases}$	70	30
	ϵ units crystals (A)	ϵ†	A = 100		
				70	30
T_2	30 units melt	67	$\begin{cases} A = 55 \\ B = 45 \end{cases}$	37	30
	15 units crystals (A)	33	A = 100	33	
	45			70	30
T_3	30 units melt	48	$\begin{cases} A = 37.5 \\ B = 62.5 \end{cases}$	18	30
	32.5 units crystals (A)	52	A = 100	52	
	62.5			70	30
T_4^+	30 units melt	43	$\begin{cases} A = 30 \\ B = 70 \end{cases}$	13	30
	40 units crystals (A)	57	A = 100	57	
	70			70	30
	Melt solidifies to form eutectic microstructure				
T_4^-	30 units eutectic xtals	43	$\begin{cases} A = 30 \\ B = 70 \end{cases}$	13	30
	40 units crystals (A)	57	A = 100	57	
	70			70	30

†ϵ = infinitesimal amount of

3.1. The Binary Eutectic

The tracing of the cooling path of melt X in Fig. 3.3 illustrated the methods of analysis of two-component systems. The temperature of the freezing point of pure A crystals is shown at M. Additions of component B to component A cause this freezing point to decrease, and this lowering of the freezing point is expressed by the downward slope of the liquidus line ME. In the course of crystallization of melt X, the crystals of pure A begin to appear at point X_1. Further cooling causes more A crystals to separate from the melt. This separation of primary A crystals from the melt leaves the melt richer in component B, and the composition of the melt changes during crystallization as indicated by the liquidus line from point X_1 to point E. At point E, the last liquid remaining would crystallize on cooling to form a crystalline mixture of both A and B. This reaction involving a liquid changing to two crystalline phases is called the eutectic reaction. Point E is the eutectic point. At this point the melt is of such a composition that this eutectic reaction takes place as

melt $E \leftrightarrows A$ crystals + B crystals

The eutectic reaction is an isothermal three-phase reaction in which a liquid is in equilibrium with two crystalline phases.

At point N_1 in Fig. 3.3, a melt of composition Y would begin to crystallize in a manner similar to the crystallization of melt X_1 except the material first separating from melt Y would be B crystals. Crystallization of B crystals from the melt would leave the melt richer in component A, and the composition of the melt during cooling would change as indicated by the liquidus line from point N_1 to E. The last liquid to crystallize would again be the melt at the eutectic point E. The eutectic point formed by the intersection of the two liquidus lines represents the composition of the lowest melting mixture in this system.

If a microscope were arranged so as to view melt X at the various stages of crystallization, the phases might appear as shown schematically in Fig. 3.6. At point 1, crystals begin to appear, but the system at point 1 consists essentially of all liquid. At point 2, the schematic representation of phases shows the growth of crystals so that about one-third of the system is composed of crystals. Further development of crystals takes place on cooling and a greater amount of the crystalline phase appears at point 3 and point 4. At

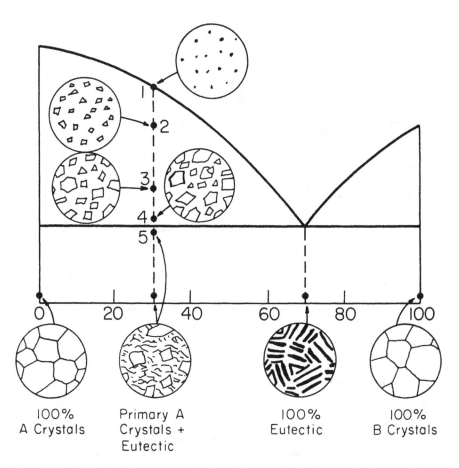

Fig. 3.6.   Schematic representation of crystals during and after cooling.

point 4, the melt composition is at the eutectic composition, E. Cooling from point 4 to point 5 changes the eutectic melt composition to a eutectic mixture of crystals of A and B. Frequently, the analysis of phases present after cooling is expressed with reference to the amounts of primary crystals and eutectic mixture present. The sketches of microscopic views of several other crystallized melts are shown at the bottom of Fig. 3.6. The relative amounts of eutectic crystals and primary crystals are shown.

In a binary system a melt of eutectic proportions solidifies to form a characteristic *microstructure** in which the two phases can always be distinguished. Sometimes the eutectic structure consists of alternate bands of the two phases. In some cases one phase may be rodlike or columnar and is surrounded by a continuous matrix of the second phase. There are many different kinds of eutectic structures; probably the most commonly observed is that described as lamellar—alternate layers or lamellae of the two crystalline phases (Fig. 3.7). These phases grow simultaneously from the supercooled melt in such a manner that each lamella is directly in contact with the melt at its advancing surface. The sequence of events may occur as shown in Fig. 3.8. The phase A nucleates first and begins to grow. As the crystal of A grows, it removes atoms of A from the melt and causes the surrounding liquid to become richer in B, which favors the nucleation of phase B. Phase B forms nuclei at the interface between A and the melt and begins to grow. The region

*The term "microstructure" is used to describe the distribution of phases on a microscopic scale.

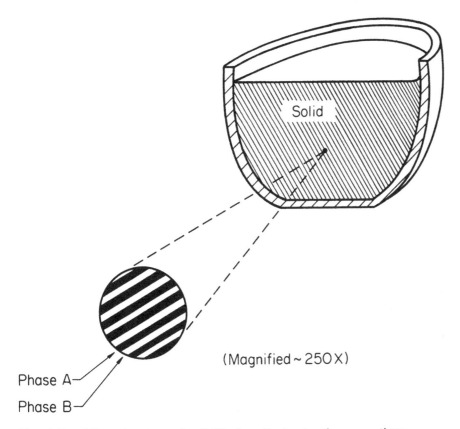

Fig. 3.7. Microstructure of solidified melt of eutectic proportions.

surrounding the growing B phase subsequently becomes rich in A, which then nucleates at the interface between B and the melt and begins to grow, thus repeating the cycle and thereby generating a lamellar structure. As the lamellae grow simultaneously into the melt, counterdiffusion of A and B atoms takes place at the advancing melt–solid interface. The rate at which the eutectic structure advances into the melt is governed by the degree of super-cooling at the interface and diffusion rates of the atomic species associated with the two kinds of lamellae.

Recently there has been a great deal of interest in understanding and controlling the way in which a eutectic solidifies because the controlled growth offers the possibility of producing materials with new and useful properties. The electrical, magnetic, optical, mechanical, and thermal prop-erties can be altered and improved by controlling the microstructure via directional solidification of melts or aligned crystallization of glasses. Tough

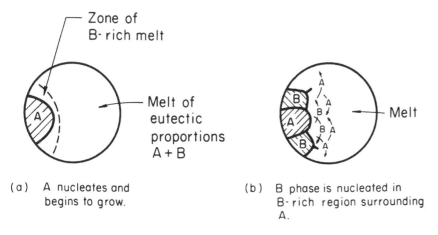

(a) A nucleates and begins to grow.

(b) B phase is nucleated in B-rich region surrounding A.

Fig. 3.8. Formation of eutectic microstructure.

composite materials can be produced by using the controlled eutectic approach to grow parallel arrays of strong fibers in appropriate matrices.

3.2. Intermediate Compounds

Compounds composed of two or more components frequently occur in a system and are referred to as *intermediate compounds*. In binary systems these compounds are composed of various ratios of the two components of the system. In Fig. 3.9 the compound AB_2 is formed of the components A and B in the mole ratio of 1/2; that is, there is one A atom for every two B atoms in the compound.

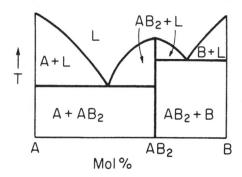

Fig. 3.9. Binary system with intermediate compound.

These intermediate compounds are classified with respect to their behavior during melting. The compound AB_2 is called a *congruently melting* compound because it melts directly to a liquid of chemical composition identical to that of the crystalline compound. Congruently melting compounds divide the system into separate, smaller binary systems. In the system in Fig. 3.9, the compound AB_2 divides the system A-B into the systems AB_2-A and AB_2-B, both of which are simple eutectic systems. In analyzing the reactions which may occur on cooling a melt to final solidification, it is customary to give the composition of the liquid phase in terms of the components of the system. The compositions of crystalline phases are always expressed in terms of the crystalline phases which are present or, if appropriate, in terms of the primary crystals and the eutectic mixture. A cooling study in the system MnO-Al_2O_3 (Fig. 3.10) will serve to illustrate the conventions which are in common usage in the ceramic field.

The composition axis may be given in weight percent or in mole percent. If there is a large difference between the formula weights of the components of the system, the phase diagram with its composition axis given in weight percent will look somewhat different from one using mole percent for the composition axis; the intermediate compounds will be located at different positions. In the system MnO-Al_2O_3 the position of the intermediate compound $MnO \cdot Al_2O_3$ on the weight percent axis is determined as follows:

Atomic wt of Mn = 54.94

Atomic wt of O = 16.00

Formula wt of MnO = 54.94 + 16 = 70.94

Formula wt of Al_2O_3 = 2(26.98) + 3(16) = 101.96

$$\text{wt\% } Al_2O_3 = \frac{101.96}{172.90} \times 100 = 59\%$$

$$\text{wt\% } MnO = \frac{70.94}{172.90} \times 100 = 41\%$$

Thus, the compound $MnO \cdot Al_2O_3$ is located at 59% Al_2O_3 as shown on the diagram. If the composition axis were given in mole percent, the compound $MnO \cdot Al_2O_3$ would be located at 50% Al_2O_3.

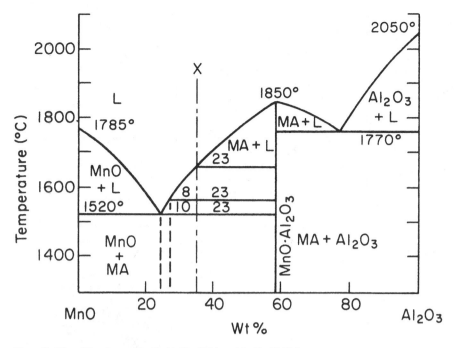

Fig. 3.10.   System MnO-Al₂O₃ (MA = MnO·Al₂O₃).

The vertical line labeled X on the MnO-Al$_2$O$_3$ diagram represents a line of constant composition and is referred to as an *isopleth*. The determination of the reactions which occur on cooling a sample of constant total composition is referred to as an *isoplethal study*. The calculations for such a study during the cooling of a melt of 35 wt% Al$_2$O$_3$ and 65 wt% MnO are recorded in Table 3.2. Note that the terms 1520$^+$ and 1520$^-$ are used to denote

Table 3.2.   Isoplethal Study in System MnO-Al$_2$O$_3$ for Composition 35% Al$_2$O$_3$, 65% MnO

Temp (°C)	Proportions (phases)	% (Phases)	% Composition of phases (in terms of components)	Analysis MnO	Analysis Al₂O₃
1670	23 units melt	100	MnO = 65 / Al₂O₃ = 35	65	35
	ε units crystals (MA)	ε	MnO = 41 / Al₂O₃ = 59		
				65	35
1560	23 units melt	74	MnO = 73 / Al₂O₃ = 27	54	20
	8 units crystals (MA)	26	MnO = 41 / Al₂O₃ = 59	11	15
	31			65	35
1520$^+$	23 units melt	70	MnO = 75 / Al₂O₃ = 25	52.4	17.5
	10 units crystals (MA)	30	MnO = 41 / Al₂O₃ = 59	12.5	17.5
	33			65	35
Melt solidifies to form eutectic microstructure					
1520$^-$	23 units eutectic xtals	70	MnO = 57.5 / MnO·Al₂O₃ = 42.5	40.3 12.2	17.5
	10 units crystals (MA)	30	MnO = 41 / Al₂O₃ = 59	12.5	17.5
	33			65	35

29

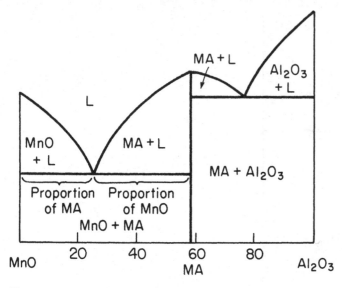

Fig. 3.11. Use of the lever rule to determine eutectic proportions.

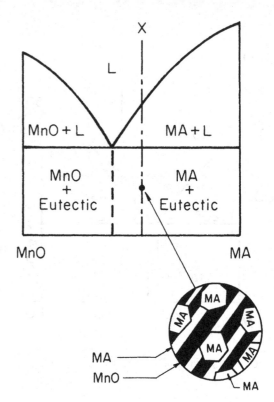

Fig. 3.12. Sketch of microstructure of composition X cooled to temperature indicated.

temperatures slightly above and slightly below the eutectic temperature. At 1520^- the eutectic reaction is complete and the 30% melt which was present at 1520^+ has solidified in eutectic proportions. The eutectic reaction is

$$MnO + MnO \cdot Al_2O_3 \overset{1520°C}{\leftrightarrows} melt$$

The proportions of MnO and $MnO \cdot Al_2O_3$ which make up the eutectic are determined from the position of the eutectic point with respect to its constituents MnO and $MnO \cdot Al_2O_3$ (Fig. 3.11).

The two-phase region $MnO + MnO \cdot Al_2O_3$ may be thought of as subdivided into two smaller areas with the eutectic composition as a common boundary as shown in Fig. 3.12. This is the treatment used in Table 3.2 and is preferred because it takes cognizance of the microstructure of the cooled sample. In this particular case, the microstructure might look as shown in the sketch of Fig. 3.12.

Unlike the intermediate compounds that melt congruently, the incongruently melting compounds do not melt to a liquid of the same composition but change to a different liquid and another crystalline phase. Figure 3.13 shows the system V_2O_5-Cr_2O_3 in which the intermediate compound $Cr_2O_3 \cdot V_2O_5$ transforms, on heating, to Cr_2O_3 and melt. Along line PN on the diagram, three phases are in equilibrium as shown by the reaction:

$$Cr_2O_3 \cdot V_2O_5 \overset{810°C}{\leftrightarrows} Cr_2O_3 + melt$$

This reaction of incongruent melting is often called a *peritectic reaction*. A peritectic reaction is an isothermal three-phase reaction in which a crystalline phase is in equilibrium with another crystalline phase and a liquid. The point P is called the peritectic point. Along the line PN there are three phases in equilibrium: $Cr_2O_3 \cdot V_2O_5$, Cr_2O_3, and liquid of the composition at point P.

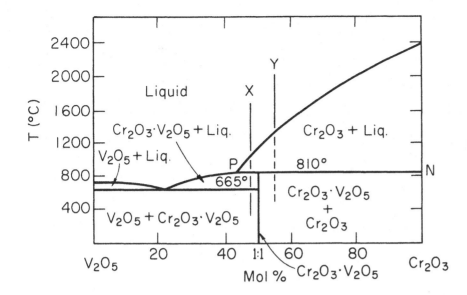

Fig. 3.13. System V_2O_5-Cr_2O_3.

Consider an isoplethal study for the composition labeled X in Fig. 3.13. When melt X is cooled to the temperature corresponding to the intersection of the isopleth with the liquidus line, crystals of Cr_2O_3 will begin to precipitate from the melt. As cooling continues, the quantity of Cr_2O_3 increases until the peritectic temperature, 810 °C, is reached. As the sample is cooled slightly below 810 °C, the peritectic reaction occurs:

$$Cr_2O_3 + melt \rightarrow Cr_2O_3 \cdot V_2O_5$$

All of the Cr_2O_3 crystals which were present before the reaction took place have reacted with the melt and a new crystalline phase, $Cr_2O_3 \cdot V_2O_5$, has formed. The *composition* of the melt has not changed, but the *quantity* of melt has been reduced. The elimination of the Cr_2O_3 is referred to as "resorption." One might visualize the resorption reaction as dissolution of the Cr_2O_3 with the simultaneous formation of the new crystalline phase. The exact mechanism for the reaction can only be postulated and likely would be different in different systems. The sequence of reactions or structural rearrangements which must occur will be accomplished in a manner which is energetically most favorable, that is, with the least disruption of chemical bonds. It may, in the case of Cr_2O_3, involve only partial solution of the Cr_2O_3 phase. If cooling is too rapid to permit the attainment of equilibrium, we would expect to see a microstructure in which the reaction was only partially completed. This structure is shown in the sketch of Fig. 3.14, where the $Cr_2O_3 \cdot V_2O_5$ phase has precipitated on the surface of the Cr_2O_3 particle. In order to complete the peritectic reaction, diffusion of the atoms or ions of Cr, V, and O must take place in the solid state. This latter reaction is slow compared to diffusion in the liquid state; thus, the approach to equilibrium will be slow and the "coring" effect shown in the sketch is likely to occur.

Cooling a melt of composition Y (Fig. 3.13) to below the peritectic temperature would result in crystals of both Cr_2O_3 and $Cr_2O_3 \cdot V_2O_5$. There is an insufficient quantity of liquid to react with all of the Cr_2O_3 which is present above the peritectic temperature; consequently, this excess of crystalline Cr_2O_3 remains as one of the solidification products after the peritectic reaction is completed.

Finally, if the melt composition corresponds to that of the compound $Cr_2O_3 \cdot V_2O_5$, the peritectic reaction on cooling will yield but one phase, $Cr_2O_3 \cdot V_2O_5$.

Decomposition

Intermediate compounds may also decompose on heating or cooling, forming two other crystalline phases. In Fig. 3.15 the intermediate compound

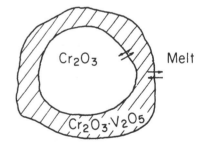

Fig. 3.14. "Coring" effect which may occur during a peritectic reaction, Cr_2O_3 + melt→$Cr_2O_3 \cdot V_2O_5$, if cooling is too rapid for attainment of equilibrium.

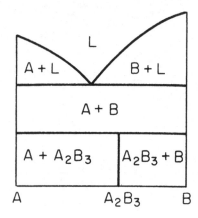

Fig. 3.15. Decomposition of A_2B_3.

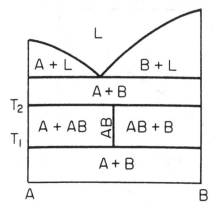

Fig. 3.16. Compound AB is stable between temperatures T_1 and T_2 only.

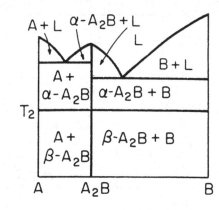

Fig. 3.17. Compound A_2B undergoes α-β phase transformation at T_2.

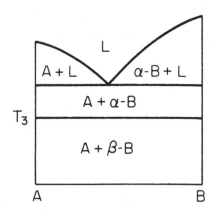

Fig. 3.18. Component B transforms from the α to β form at T_3 on cooling.

A_2B_3 decomposes on heating to form A and B. In Fig. 3.16 the compound AB is stable only between temperatures T_1 and T_2.

Phase transformations may occur during the heating or cooling of intermediate compounds. In Fig. 3.17 the compound β-A_2B transforms to α-A_2B at temperature T_2. In Fig. 3.18 the component B undergoes a phase transformation at temperature T_3.

3.3. Solid Solution

When a melt composed of two or more kinds of atoms is cooled to solidification, there are three different reactions which may occur. These reactions are illustrated schematically in Fig. 3.19, where a melt of two different kinds of atoms is shown on the left-hand side. The atomic species appearing in greatest concentration is usually called the solvent and the species occurring in the least concentration is termed the solute. If during solidification the inclusion of solute atoms in the crystalline structure of the solvent would lead to an increase in the energy of the lattice, the solute atoms may be rejected and two distinct crystalline phases will be formed. This reaction is shown in part I of Fig 3.19.

If, however, the inclusion of the solute atoms in the solvent lattice in some *ordered* way (involving a simple ratio of the two kinds of atoms) results in a lower energy system, then a compound will be formed as shown in part II of Fig. 3.19.

In the third possibility, solute atoms may fit into the solvent lattice in a random manner to form a single, crystalline phase. This reaction results in the formation of a solid solution (Fig. 3.19, part III).

The term *solid solution* is used both as a verb and as a noun. When used as a noun, it may be defined as a single crystalline phase which may be varied in composition within finite limits without the appearance of an additional phase. Used as a verb, solid solution refers to the means by which an atomic species may occur in a compound in other than stoichiometric amounts.

The solute atom may fit into the host lattice in two distinct ways:

1. Substitutional solid solution. The solute atoms fill lattice positions normally occupied by solvent atoms. The solute atoms are, in effect, "substituting" for solvent atoms (Figs. 3.20(a)).

2. Interstitial solid solution. The solute atoms fit into the voids or interstices formed by the solvent or host atoms (Fig. 3.20(b)).

The extent of substitutional solid solution is determined by several factors:

Size. If the solute atom is of approximately the same size as the host atom, extensive substitution may occur. If, however, the size disparity is greater than about ±15%, the substitution is usually limited to a few percent.

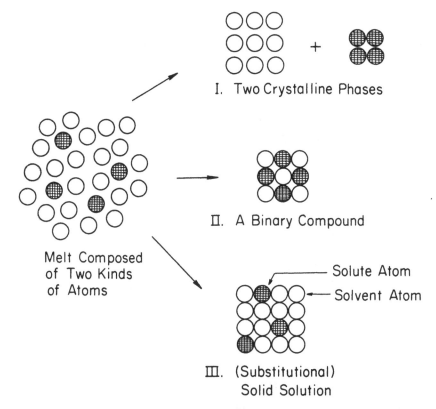

I. Two Crystalline Phases

II. A Binary Compound

Solute Atom
Solvent Atom

Melt Composed
of Two Kinds
of Atoms

III. (Substitutional)
Solid Solution

Fig. 3.19. Possible reactions during solidification of a melt.

Chemical affinity. If there is a strong chemical affinity between the two kinds of atoms, a more stable configuration (lower free energy) may be attained by the formation of a compound. In the latter case, solid solubility would be restricted to small amounts.

Valency. If the solute atom differs in valence from the solvent atom, solid solution will be limited because structural changes will be required in order to preserve the overall electrical neutrality. For example, substitution of a trivalent ion for a divalent ion can be accommodated if an occasional atom site is left vacant.

Structure type. The occurrence of complete solubility between end-members requires that the crystal structures of the two members be alike.

Examples in Ceramic Systems

Solid solution in ceramic materials usually involves systems containing more than two kinds of atoms. For example, solid solution between two oxides such as NiO and MgO involves three kinds of atoms: Mg, Ni, and O. Both NiO and MgO have cubic crystal structures, Ni and Mg have equal

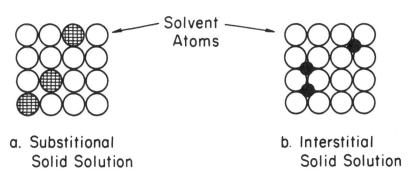

Solvent Atoms

a. Substitional
Solid Solution

b. Interstitial
Solid Solution

Fig. 3.20.   Types of solid solution.

Table 3.3. Ionic Radii* (for Coordination Number of 6)

Ion	Radius (Å)	Range (±15%)
Na⁺	0.95	0.81–1.09
K⁺	1.33	1.13–1.53
Li⁺	0.60	0.51–0.69
Ni²⁺	0.69	0.59–0.79
Mg²⁺	0.65	0.55–0.75
Fe²⁺	0.75	0.64–0.86
Ca²⁺	0.99	0.84–1.14
Ba²⁺	1.35	1.15–1.55
Sr²⁺	1.13	0.96–1.30
O²⁻	1.40	1.19–1.61
Al³⁺	0.50	0.43–0.58
Si⁴⁺	0.41	0.35–0.46
Ti⁴⁺	0.68	0.58–0.78
Zr⁴⁺	0.80	0.68–0.92
Ce⁴⁺	1.01	0.86–1.26
Pb⁴⁺	0.82	0.71–0.97

*After L. Pauling, "The Nature of the Chemical Bond," 3rd ed. Cornell University Press, Ithaca, NY, 1960.

valences, and the ionic radius of Mg is within ±15% that of Ni (see Table 3.3). One would expect extensive solid solution between MgO and NiO, and this is what occurs. Figure 3.21 shows the system MgO-NiO; complete solid solution exists between the end-members.

The Ca ion exceeds the ±15% approximation, and as seen in Fig. 3.22, solid solution between CaO and MgO is limited.

The limitation on the extent of interstitial solid solution which may occur is that the apparent diameter of the solute atom is smaller than 0.6 times that of the solvent atom.

When a melt of composition X is cooled in the system NiO-MgO (Fig. 3.23), the first solid appears at a temperature corresponding to the intersection of the isopleth with the liquidus line. This infinitesimal bit of solid solution has the composition [MgO] = 63% and [NiO] = 37%.

Further cooling of the sample to approximately 2350°C results in an increase in the amount of solid solution and a change in the composition of both the solid solution and the liquid phase. These compositions are given by

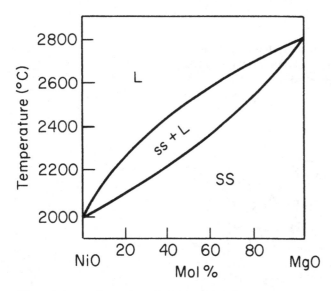

Fig. 3.21. System NiO-MgO (after H. von Wartenberg and E. Prophet, *Z. Anorg. Allg. Chem.*, **208**, 379 (1932)).

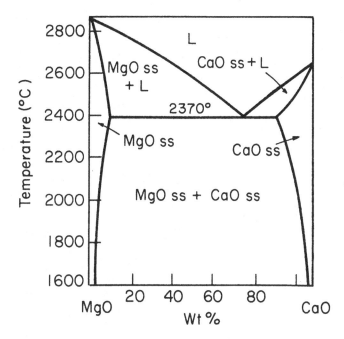

Fig. 3.22. System MgO-CaO (after R. C. Doman, J. B. Barr, R. N. McNally, and A. M. Alper, *J. Am. Ceram. Soc.*, **46** [7] 313–16 (1963)).

the intersections of the 2350°C tie line with the solidus and liquidus lines, respectively.

At 2250°C, the isopleth intersects the solidus line. The last trace of liquid contains approximately 20% MgO and 80% NiO. The solid portion of the sample is a solid solution of 40% MgO and 60% NiO. No additional changes occur when the sample is cooled to lower temperatures.

The changes in composition of the melt and of the solid solution during the cooling of the sample through the two-phase region are shown by the arrows in Fig. 3.23. The corresponding calculations are recorded in Table 3.4.

It should be remembered that these calculations assume equilibrium to be attained at each temperature considered. It is apparent that the first solid phase to precipitate is very rich in MgO. At the next lower temperature, the

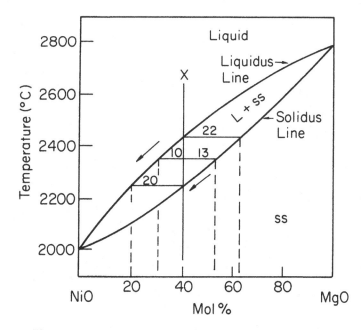

Fig. 3.23. System MgO-NiO.

35

Table 3.4. Isoplethal Study in System MgO-NiO Composition = 60% NiO, 40% MgO

Temp (°C)	Proportions (phases)	% (Phases)	% Composition of phases (in terms of components)	Analysis NiO	Analysis MgO
2430	22 units melt	100	MgO = 40 / NiO = 60	60	40
	ε units crystals (SS)	ε	MgO = 63 / NiO = 37		
				60	40
2350	13 units melt	56.5	MgO = 30 / NiO = 70	39.5	17
	10 units crystals (SS)	43.5	MgO = 53 / NiO = 47	20.5	23
	23			60.0	40
2250	ε units melt	ε	MgO = 20 / NiO = 80		
	20 units crystals (SS)	100	MgO = 40 / NiO = 60	60	40
				60	40

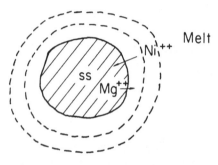

Fig. 3.24. (Mg$_x$Ni$_{1-x}$)O solid solution growing from melt. Counterdiffusion of Mg^{2+} and Ni^{2+} is required to obtain overall equilibrium composition.

composition of the solid solution is less rich in MgO. To obtain this equilibrium composition, a counterdiffusion of Mg and Ni ions must occur between the original crystal and the layers which are subsequently deposited at the lower temperatures (Fig. 3.24). Since diffusion in the solid state is relatively slow, it is to be expected that in most cases (unless cooling is extremely slow) a composition gradient will exist in the solid-solution crystals.

It is also possible to have a solid-solution series with a maximum or minimum. Figure 3.25 shows the system CaO·SiO$_2$-SrO·SiO$_2$ which contains a minimum point. The system Pb-Tl contains a maximum point (Fig. 3.25). The points labeled "I" in the two figures are called "indifferent or congruent points." At the congruent points both the liquidus and solidus curves are tangent to each other and to the isothermal line through point I. The points I are invariant by restriction. Since the composition of the liquid and solid solution at point I is the same we may set the number of components $C = 1$. Thus the phase rule $F = C - P + 1$ gives $F = 1 - 2 + 1 = 0$.

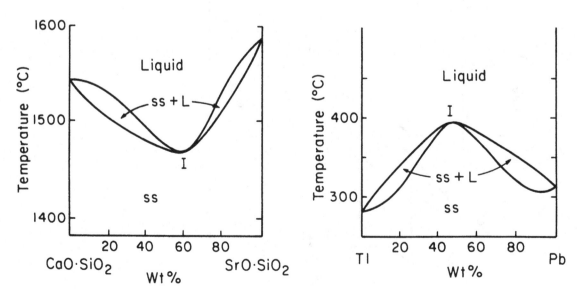

Fig. 3.25. Systems with solid solution maxima and minima.

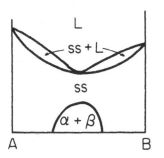 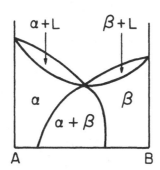 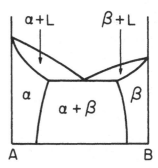

Fig. 3.26. Extension of solubility gap in solid solution phase to form eutectic system.

A solubility gap sometimes occurs in the solid-solution phase and in some cases may extend to the liquid phase, forming a eutectic system. Several variations are shown in Fig. 3.26.

Several examples of typical phase diagrams with solid solution are shown in Fig. 3.27.

It can be reasoned that every system should exhibit solid solution to a certain extent. There is a thermodynamic driving force for a pure substance to in-

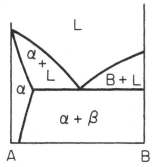 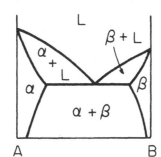 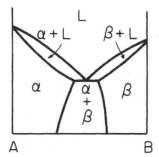

A. Solid solution of B in A to form α ss

B. Solid solution of B in A and of A in B

C. Greater degree of solid solution

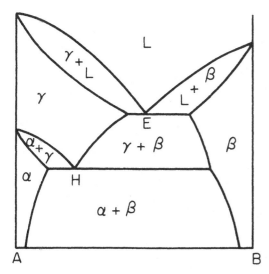

D. Point E is a eutectic point. Point H is a eutectoid point; three solid phases, α, β, and γ are in equilibrium.

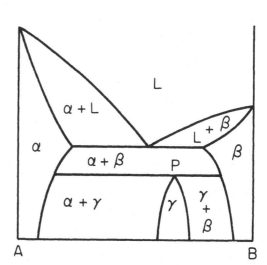

E. Point P is a peritectoid point. The peritectoid reaction is $\alpha + \beta = \gamma$

Fig. 3.27. Types of solid solutions in several phase diagram constructions.

37

clude at least a few foreign atoms in its structure because by so doing the entropy of the system is increased (see Chapter 4 for additional discussion). In many cases the extent of solid solution is so slight that it would appear on the diagram as less than the width of a line, and solid solubility can thus be considered to be negligible.

3.4. Liquid Immiscibility

Unlike gases, which are completely miscible in all proportions, liquids are not in every case completely soluble in one another. Ether and water, for example, do not completely dissolve in one another, although there is some slight solubility of water in ether and some solubility of ether in water. When two liquids do not completely dissolve in one another, the liquids separate out as two phases, a condition which is referred to as liquid immiscibility. Figure 3.28 shows an example of a system in which liquid immiscibility occurs.

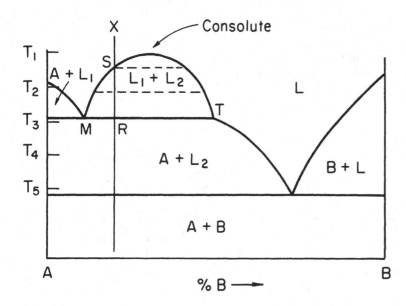

Fig. 3.28. Binary system in which liquid immiscibility occurs.

Area MST is the two-liquid region. Cooling a melt of composition X through this region would result in the appearance of two liquid phases at point S. The compositions of the two liquids are given by the intersections of the tie line with the boundary of the two-liquid region. Further cooling results in a change in the composition of each phase. At a temperature corresponding to the isothermal line MRT, the two liquids have the compositions shown by points M and T, respectively. Cooling to a temperature below MRT will cause the following reaction to occur:

$$L_1 \to A + L_2$$

Liquid L_1 changes to crystals of A and liquid of composition L_2. This reaction is often referred to as a *monotectic reaction*. It is a three-phase, isothermal reaction similar to the eutectic reaction except that the liquid changes to one solid phase and another liquid rather than changing to two solid phases as in the case of the eutectic reaction. The point M is called the *monotectic point*. All melt compositions lying between the limits M and T will undergo the monotectic reaction when cooled through the monotectic temperature. Since three phases coexist along the line MRT, the variance (as calculated from the phase rule) is zero and therefore the temperature cannot vary and the MRT must be isothermal.

When liquid–liquid separation occurs, the microstructure of the melt undergoes a change. In some instances the second liquid phase separates out

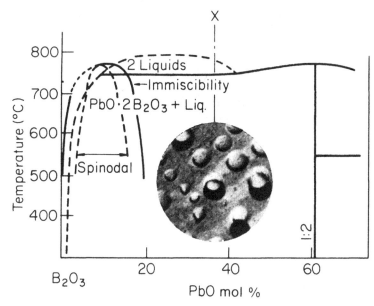

Fig. 3.29. Photomicrograph of melt X in system B_2O_3-PbO quenched to room temperature. Dashed line indicates proposed phase boundaries (after K. E. Geller and E. H. Bunting, *J. Res. Natl. Bur. Stand. (U.S.)*, **18** [5] 585, 589 (1937)). Immiscibility boundary and estimated spinodal proposed by J. H. Simmons. *J. Am. Ceram. Soc.*, **56,** 284 (1973).

as discrete droplets dispersed in a matrix of a second phase (Fig. 3.29). In other cases the boundaries between the phases will be diffuse and both phases may be continuous (e.g., spinodal decomposition).

The reasons for the separation of a liquid into two phases may be found by considering the thermodynamics of a nonideal solution. In such solutions, the attractions between like and unlike atoms are not equal; consequently, when A atoms are mixed with B atoms, there will be an enthalpy change as well as an increase in entropy due to the mixing of unlike atoms. As shown in Chapter 4, if the attraction of A atoms for B atoms is greater than that of A for A or B for B, the enthalpy of mixing is negative. If, on the other hand, the forces of attraction between like atoms are greater than between unlike atoms, there is a tendency toward clustering of like atoms or an unmixing of the solution to form two phases. The diagram of Fig. 3.30 depicts the latter state at a temperature T for a system of A and B atoms.

The free energy of a solution of A and B as a function of composition can be related to the phase diagram as shown in Fig. 3.31 for temperature T_1. The tangent line ad in the lower figure indicates the equilibrium phases at that temperature for all compositions between a and d. A melt whose composition lies between a and d and which is cooled to temperature T_1 will tend to separate into two liquids, the relative proportions of which will depend upon the position of the isopleth relative to the tie line ad in Fig. 3.31(A).

The inflection points, b and c, on the free energy curve have special significance. Melt compositions between a and b and between c and d are metastable and will remain as a single liquid when cooled to T_1 unless the second phase is somehow nucleated. The slope of the free energy curve between a and b and between c and d indicates that a very small, random fluctuation in composition would increase the free energy of the system and thus would preclude spontaneous separation into two liquids. In order to nucleate the second phase, some discontinuity, such as a container wall, a bubble, or an impurity, must be present to reduce the free energy barrier to the formation of a stable nucleus. It is thus possible to supercool such melts without the occurrence of liquid-liquid separation.

Melts having compositions between b and c are unstable at temperature T_1. The free energy-composition curve indicates that an infinitesimal fluctua-

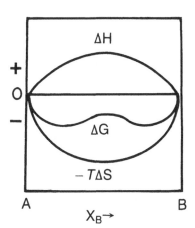

Fig. 3.30. Energy diagram for hypothetical system in which unmixing occurs.

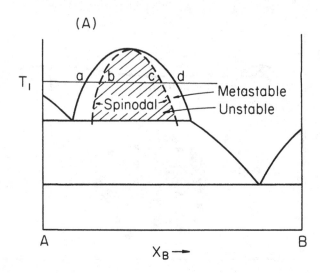

(A)

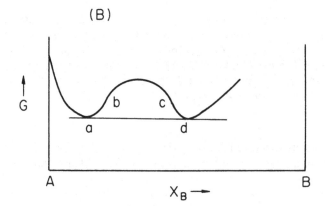

(B)

Fig. 3.31. (A) Binary phase diagram with liquid miscibility gap. (B) Free energy as a function of composition at temperature T_1 for a solution of A and B in which liquid immiscibility occurs.

tion in composition is accompanied by a decrease in the free energy of the system. Thus, a minute fluctuation in composition will cause the melt to separate into two liquid phases whose composition will change continuously with time to that given by points a and d, the equilibrium compositions at temperature T_1.

The region defined by the dashed lines is called the *spinodal*. The microstructure of melts separated by the spinodal decomposition mechanism is characterized by diffuse boundaries between the phases. Both liquid phases are continuous and are independently interconnected during the initial stages of separation. In general, melts whose compositions fall within the spinodal during cooling cannot be easily supercooled without the appearance of a second phase.

Liquid–liquid separation has important applications in ceramics. In the area of glass-ceramics (sometimes called recrystallized glasses), nucleation of the crystalline phase may occur at the energetically favorable interface between the liquid phases; thus, liquid–liquid separation provides an effective means for nucleating the crystallization process within a glass.

Fused silica[†] (96% silica glass) shapes are produced by selectively leaching one phase of a spinodally decomposed borosilicate glass. The phase lower in silica content is leached away and a subsequent heat treatment consolidates the structure to form dense, high-silica glassware. It would not be economically feasible to shape the 96% silica glass by conventional methods because of the very high temperatures required to work this extremely viscous glass.

_____

[†]Vycor, trade name of Corning Glass Works, Corning, New York.

3.5. Structural Considerations of Liquid Immiscibility

Consideration from a crystal chemistry point of view has contributed to our understanding of liquid immiscibility in glass-forming systems and has had some success in predicting limits of composition for the two-liquid region.

Warren and Pincus pictured the atomic bonding in a glass as largely ionic and calculated the bonding energy from the equation:

$$E = (-z_1 z_2 e^2)/R_{12}$$

E = bonding energy between an ion pair
z_1 and z_2 = valences of the respective ions
e = charge on the electron
R_{12} = separation between ions

The Si–O bond in a silicate glass would be extremely stable because of the high charge and small ionic radius of the silicon atom ($Si = 4+$, $O = 2-$, $R_{12} = 1.62$ Å). By contrast, the Na–O bond is relatively weak ($Na = 1+$, $O = 2-$, $R_{12} = 2.35$ Å).

Oxides are broadly classified as network formers or network modifiers, depending upon their bonding arrangements in the glass. Network formers are those oxides which have large cation–oxygen bond energies, such as SiO_2, B_2O_3, and P_2O_5. In a glass these oxides form a continuous three-dimensional network by sharing oxygen ions. The modifier oxides, such as Na_2O, K_2O, CaO, which have much smaller cation–oxygen bond energies modify the network by contributing oxygen ions which reduce the number of shared oxygens in the network and thus cause a more open structure. The modifier cations take up positions in the interstices of the network and satisfy their charge by coordinating with several partially bonded oxygens. A schematic sketch of this arrangement is shown in Fig. 3.32. The competition of the network cations and the modifier cations for the available oxygen ions leads to a compromise structure which may involve the separation of the melt into two liquids.

The system Na_2O-SiO_2, which exhibits no stable liquid immiscibility (metastable subliquidus glass-in-glass immiscibility is obtainable in this system), is analyzed by Warren and Pincus in the following way:

"Each silicon in the melt is striving to keep itself continually bonded to four oxygens in spite of occasional breaking of bonds by the thermal agitation. This is more easily accomplished with a single

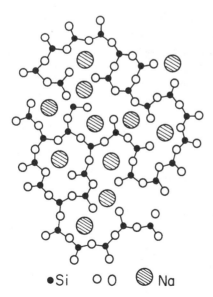

•Si   ○ O   ⊘ Na

Fig. 3.32. Schematic representation of atomic arrangement in soda-silica glass.

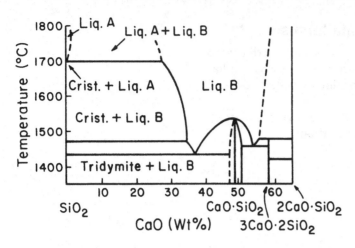

Fig. 3.33. System SiO_2-$2CaO \cdot SiO_2$ (after J. W. Greig, *Am. J. Sci.*, **13** [73] 1–44; [74] 133–54 (1927)).

phase for the soda–silica melt because the number of oxygens per silicon is higher and it is not necessary for every oxygen to be bonded between two silicons in order that each silicon shall be bonded to four oxygens.''

''Working in the other direction is the desire for each sodium ion to have a proper surrounding of oxygens. The oxygens, bonded only to one silicon, are the only unsaturated oxygens, and the Na^+ ions will try to surround themselves by these unsaturated oxygens. With low soda content, the number of Na^+ ions and unsaturated oxygens is small, and if they are uniformly distributed throughout the glass, they will be widely separated. It would be easier for each Na^+ ion to be in contact with several unsaturated oxygens if a phase of higher soda content were to segregate. In a soda–silica melt with low soda content, therefore, are two opposing tendencies. The continual bonding and rebonding of the silicon with all available oxygen in the melt tends to hold all of the oxygen in a single phase, whereas the attraction between Na^+ ions and unsaturated oxygens tends to segregate a phase of higher soda content. Because the sodium ion is monovalent and the sodium–oxygen bond is rather weak, the second tendency is relatively weak. The first effect predominates, and soda and silica, accordingly, are completely miscible.''

In the system CaO-SiO_2 (Fig. 3.33), where the Ca–O bond is considerably stronger than the Na–O bond, stable liquid–liquid separation occurs. The attraction of the calcium ion for oxygen is of sufficient magnitude to cause the melt to separate into two liquids: one liquid containing a small percentage of Ca and the other a relatively large percentage. The calcia-rich liquid has a more flexible network owing to the greater number of nonbridg-

Table 3.5. Ratio of Cation Valence to Radius

	R	z	z/R	Type of liquidus curve
Cs	1.65	1	0.61	
Rb	1.49	1	0.67	Nearly straight
K	1.33	1	0.75	
Na	0.98	1	1.02	
Li	0.78	1	1.28	S-shaped
Ba	1.43	2	1.40	
Sr	1.27	2	1.57	
Ca	1.06	2	1.89	Immiscibility
Mg	0.78	2	2.56	

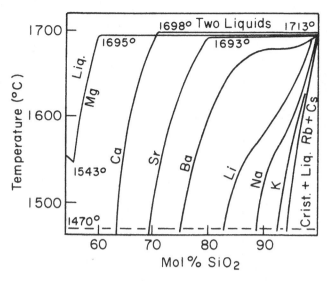

Fig. 3.34. Liquidus curves for alkali and alkaline earth oxide-silica systems (after F. C. Kracek, *J. Am. Chem. Soc.*, **52** [4] 1436–42 (1930)).

ing oxygen ions, and consequently, the calcium ion can be more easily surrounded with oxygen ions in order to satisfy its charge.

Arranging the common modifier ions in order of increasing z/R ratio (Table 3.5) shows a correlation of the tendency toward immiscibility with the shape of the liquidus curves in silicate systems (Fig. 3.34). The greater the strength of the oxygen–modifier bond, the greater is the tendency to separate into two liquids.

3.6. The System Al_2O_3-SiO_2

A relatively "simple" binary system which is of considerable importance in studying ceramic materials is the system Al_2O_3-SiO_2.

This system also illustrates some of the problems in satisfactorily determining and evaluating experimental data needed to assemble the phase diagram. Figures 3.35 and 3.36 show the phase equilibrium curves obtained

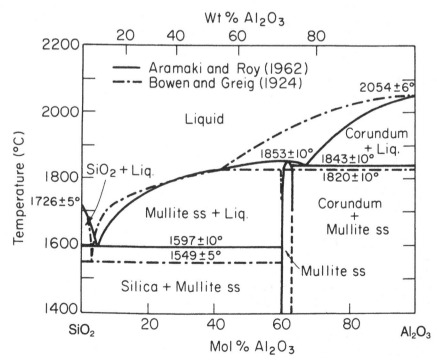

Fig. 3.35. System Al_2O_3-SiO_2 (after N. L. Bowen and J. W. Greig, *J. Am. Ceram. Soc.*, **7** [4] 238–54 (1924); and S. Aramaki and R. Roy, *ibid.*, **42** [12] 644–45 (1959), **45** [5] 229–42 (1962)).

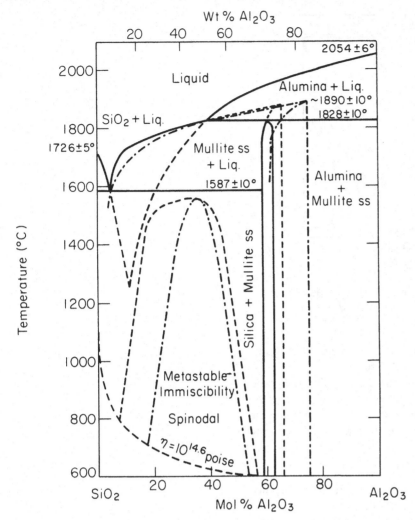

Fig. 3.36. System Al$_2$O$_3$-SiO$_2$ (after I. A. Aksay and J. A. Pask, *J. Am. Ceram. Soc.*, **58** [11–12] 507–12 (1975); and S. H. Risbud and J. A. Pask, *ibid.*, **60** [9–10] 418–24 (1977)). Calculated metastable immiscibility region shown approximately similar to that proposed in experimental work of J. F. MacDowell and G. H. Beall, *J. Am. Ceram. Soc.* **52** [1] 17–25 (1969).

by several different experimental and thermodynamic studies.

Bowen and Greig first obtained a systematic phase diagram for this system showing the compound mullite to melt incongruently at ≈1828°C. Aramaki and Roy on the other hand proposed a revision which shows a congruently melting behavior for mullite at ≈1850°C. The stable phase equilibrium diagram proposed by Aksay and Pask is in general agreement with Bowen and Greig's version. The point of including these conflicting sets of phase diagrams for the Al$_2$O$_3$-SiO$_2$ system is to alert the student to the fact that one must use a particular diagram with awareness of the purpose and experimental methods by which the original phase diagram was obtained.

Notwithstanding the problems in obtaining full agreement on the nature of phase equilibria in the system Al$_2$O$_3$-SiO$_2$, it is clear that knowledge of phases in this system is essential to a wide variety of ceramic and glass products. SiO$_2$ is the basis for the glass manufacturing industry. Al$_2$O$_3$ is used in the production of electronic substrates, spark plugs, cutting tools, and lamp envelopes. Mullite and many other compositions in this system are useful in high-temperature refractories.

Refractories used in steelmaking include silica brick which usually contain less than 1% Al$_2$O$_3$. Examination of the liquidus curve (Fig. 3.35) shows that a liquid phase begins to form at 1595°C and that a few percent of Al$_2$O$_3$ would markedly increase the proportion of liquid in the brick at temperatures between 1595°C and the melting point of SiO$_2$. The increase in the amount of

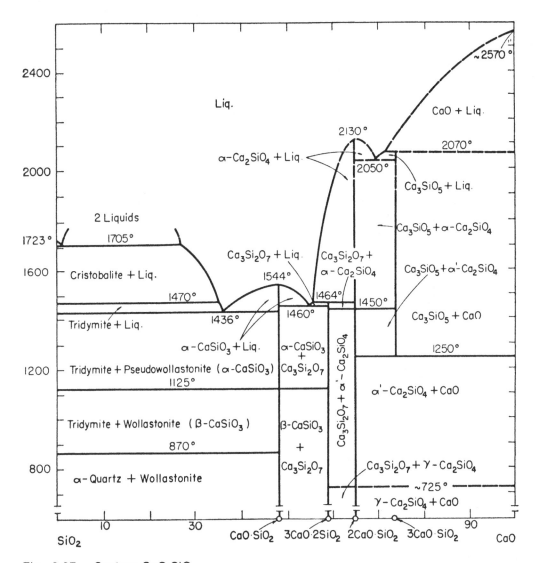

Fig. 3.37. System $CaO\text{-}SiO_2$.

liquid phase reduces the load-bearing capacity of the brick, and thus the maximum safe operating temperature is decreased by the presence of the Al_2O_3.

Fireclay refractories cover the range from about 40% Al_2O_3 to 90% Al_2O_3; the composition selected is dependent upon the particular application.

Mullite is an important refractory with a very high melting point (1850°C). Mullite can accommodate a small excess of Al_2O_3 in solid solution, as indicated on the diagram. Mixtures of mullite and Al_2O_3 form no liquid until heated above 1840°C. Mullite is used in steelmaking furnaces and in glass-melting furnaces and has also found application as an ingredient in the friction materials used in aircraft braking systems.

In addition to many refractory applications, high-purity Al_2O_3 finds use in the electronics industry because of its excellent electrical characteristics when used as an insulator in power circuits or as a substrate material in hybrid circuits for computer chips and other modern applications.

3.7. The System $CaO \cdot SiO_2$

The system $CaO\text{-}SiO_2$ illustrates many of the reactions which have been discussed in previous sections. As shown in Fig. 3.37, there are four intermediate compounds. $CaO \cdot SiO_2$ undergoes an $\beta \rightarrow \alpha$ transformation at 1125°C and melts congruently at 1544°C. The compound $3CaO \cdot 2SiO_2$ melts incongruently at 1464°C to form $\alpha\text{-}2CaO \cdot SiO_2$ and liquid. When heated to 725°C, $2CaO \cdot SiO_2$ transforms from the γ to the α' form and at 1450°C transforms from α' to α before melting congruently at 2130°C. The compound $3CaO \cdot SiO_2$ is stable between 1250°C and 2070°C. At the latter temperature it melts incongruently to form CaO and liquid.

Two liquids are in equilibrium above 1705°C for compositions containing approximately 2–28% CaO. There is no apparent solid solution between any of the phases in the system.

Problems for Chapter 3

3.1. In the binary system of Fig. 3.38, determine the following:
   a. The temperature at which 30 wt% of B crystals are in equilibrium with a liquid composed of 60 wt% B and 40 wt% A.
   b. The composition of the sample described in part a.
   c. Show by means of a sketch how the microstructure of this sample would look at the temperature determined in part a.
   d. Show by means of a sketch the appearance of the microstructure for this sample if it were cooled to 500°C.

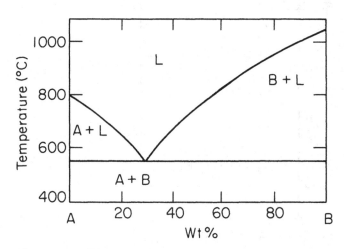

Fig. 3.38.   Binary system A-B.

3.2. System V_2O_5-NiO (Fig. 3.39).
   a. Label all areas in the diagram.
   b. Make an isoplethal study in the system on cooling a melt of 45

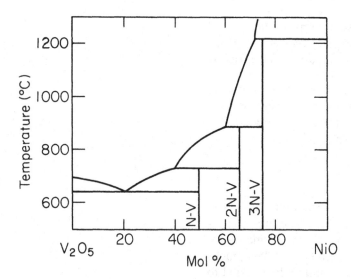

Fig. 3.39.   System V_2O_5-NiO (after C. Brisi, *Ann. Chim. (Rome)*, **47** [7–8] 806 (1957)).

mol% NiO and 55 mol% V_2O_5. Calculate before and after each change of phase:

(1) the percentage of each phase present.

(2) the percent of each component in each phase.

c. Make an isoplethal study in the system on cooling a melt of 60 mol% NiO and 40 mol% V_2O_5.

3.3. *System R-G (Fig. 3.40).*

Make an isoplethal study for a melt of composition 20 wt% G and 80 wt% R.

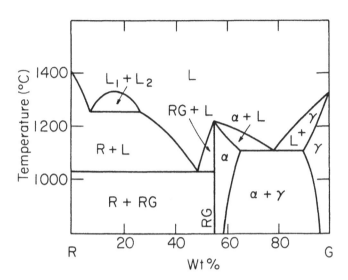

Fig. 3.40.   System R-G.

3.4. *System R-G (Fig. 3.40).*

A melt of 70 wt% G and 30 wt% R is cooled to 900°C. Calculate the percentage and the composition of each phase present.

3.5. *System NiO-MgO (Fig. 3.21).*

Make an isoplethal study for a melt of composition 70 mol% MgO and 30 mol% NiO.

Bibliography and Supplementary Reading

I. A. Aksay and J. A. Pask, *J. Am. Ceram. Soc.*, **58** [11–12] 507 (1975).

S. Aramaki and R. Roy, *J. Am. Ceram. Soc.*, **42** [12] 644 (1959); **45** [5] 229 (1962).

N. L. Bowen and J. W. Greig, *J. Am. Ceram. Soc.*, **7** [4] 238 and **7** [5] 410 (1924).

A. Burdese, *Ann. Chim. (Paris)*, **47**, 801 (1957).

R. F. Davis and J. A. Pask, *J. Am. Ceram. Soc.*, **55** [10] 525 (1972).

R. C. Doman, J. B. Barr, R. N. McNally, and A. M. Alper, *J. Am. Ceram. Soc.*, **46**, 314 (1963).

R. F. Geller and E. N. Bunting, *J. Res. Natl. Bur. Std.*, **18**, 589 (1937).

J. W. Greig, *Am. J. Sci.*, **13**, 1, 133 (1927).

F. C. Kracek, *J. Am. Ceram. Soc.*, **52**, 1436 (1930).

L. Pauling; Nature of the Chemical Bond. Cornell University Press, Ithaca, New York, 1960.

M. Tomozawa; p. 71 in Treatise on Materials Science and Technology, Vol. 17; Glass II. Academic Press, New York, 1979.

B. E. Warren and A. G. Pincus, *J. Am. Ceram. Soc.*, **23** [10] 301 (1940).

H. V. Wattenberg and E. Prophet, *Z. Anorg. Allg. Chem.*, **208**, 379 (1932).

J. H. Welch and W. Gutt, *J. Am. Ceram. Soc.*, **42** [1] 11 (1959).

In order to determine the phase relations of a system, it is necessary to establish the phases existing at various temperatures for selected compositions within the system. The positions of the invariant points, the liquidus lines or surfaces, and other points, lines, or surfaces at which phase reactions occur must be determined in such a manner that the data represent phase relations under stable or metastable *equilibrium* conditions.

Experimental methods, broadly categorized as static and dynamic techniques, have been customarily the popular approaches for determining phase equilibrium relations in a selected pressure–temperature–composition region. However, the use of thermodynamically calculated phase equilibria, cross-checked with a few critical experimental data, is experiencing wider popularity and interest with the ready availability of modern computers.

4.1. Experimental Methods

The methods used in the determination of phase equilibrium relations involve several different procedures for examining the phase reactions and identifying the phases present. At the high temperatures at which phase relations of ceramic materials are of interest, it has been difficult to directly examine the phases which are present.

Several factors are especially important in the experimental determination of phase equilibrium relations:

1. Raw materials should be of high purity. Foreign constituents may affect the system in such a manner that the system actually has, because of the foreign constituents, additional components. Such additional components may cause large deviations from the phase equilibrium relations that are intended to be measured. On the other hand, it is recognized too that foreign constituents may be beneficial rather than detrimental, for such materials may act as "mineralizers" or catalysts, increasing reaction rates and preventing the formation of metastable phases.

2. Equilibrium is more nearly approached when the area of contact of raw materials is greater. Thus, fine-particled materials intimately mixed and in contact aid in the attainment of equilibrium.

3. The time at any temperature must be sufficiently long in order to permit completeness of reactions.

4. The presence of liquid hastens reactions; the less viscous the liquid, the faster are the reactions rates.

5. The presence of viscous liquids that tend to supercool may not be desired in many cases, and a more expedient approach to equilibrium conditions may often be effected by solid-state reactions.

In the manufacture of ceramic wares, equilibrium conditions are very nearly attained with certain materials because these above-mentioned conditions are met in the firing process. With many other ceramic materials or processes, however, equilibrium conditions may be far from being met, nor, in many cases, are such conditions desired (glass, for example). Metastability and the associated phase relations under metastable equilibrium conditions are thus specially meaningful for glass and ceramic systems.

Except for the techniques involving high-temperature X-ray diffraction or the hot-stage microscope which permit a direct examination of crystals at high temperatures, the techniques of studying phase equilibria are usually indirect. Phase transformations involve a change of several physical properties, and the measurement of these physical properties as a function of temperature provides a technique for studying these transformations. The following changes are often studied in determining phase equilibrium relations:

a. structural changes
   (1) internal crystalline arrangement (X-ray diffraction analysis)
   (2) external crystalline appearance, crystal shape and habit (microscope)
b. optical changes, refractive index, birefringence, and color (microscope)
c. melting or crystallization behavior (quenching method or high-temperature microscopy)
d. heat effects (cooling or heating curves)
e. electrical changes (electrical properties such as resistivity or dielectric behavior as a function of temperature)
f. volume changes (thermal expansion curves)

If a given composition is cooled or heated at a very slow, constant rate in a furnace and the temperature of the material is measured as a function of time, the large deviations from a smooth time–temperature curve are an indication of phase changes in that material. For a material that readily crystallizes from its molten liquid, cooling curves can be used quite successfully for phase studies. Figure 4.1 shows the relation of cooling curves for several melts on a hypothetical binary system. The heat of crystallization evolved when crystals form is evidenced by the change in direction of the smooth time–temperature cooling curve. On cooling curves 1, 3, and 5 a flat horizontal portion of the curve indicates that the heat of crystallization maintains the temperature constant for a certain period of time. These plateaus are characteristic for the liquid-to-crystal transformation at the freezing points and eutectic temperatures. Curves 2 and 4 also show these plateaus at the eutectic temperature and, additionally, a change in direction of the time–temperature curve at the liquidus temperature.

Normally with ceramic melts, especially the silicate melts, the viscosity of the liquid is very high. Crystallization from such high-vicosity liquids proceeds with great difficulty, and accordingly, since the attainment of equilibrium under such conditions requires excessively long periods of time, the cooling curve method of studying phase relations of ceramic materials is not often suitable.

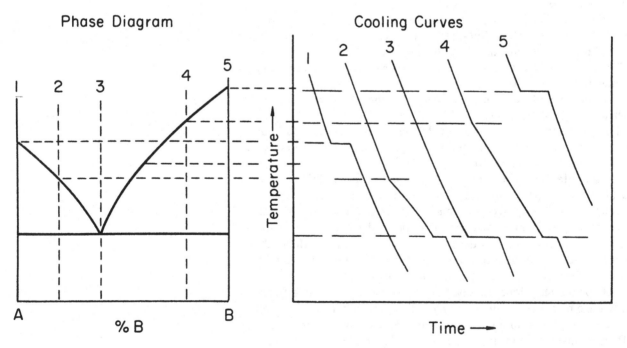

Fig. 4.1. Diagrams showing the application of the cooling curve technique for studying binary phase equilibria.

Heating curves, although not impeded by supercooling difficulties, also suffer from this sluggishness of reactions in systems containing high-viscosity liquids. The heat effects, absorption or evolution of heat, which indicate phase changes in the heating or cooling curves, are ill-defined in these systems where reactions are slow. Where viscosity is low as with liquid metals or salts, the heating or cooling curves are applicable. Heating curves (using differential thermal analysis) have been used to determine the magnitude of energy changes during crystalline phase transformations in a number of ceramic systems.

The most frequently used method of studying phase equilibria in ceramic materials has been the quenching method. In this technique, small samples are heated for sufficiently long periods of time for equilibrium to be established and then quenched in water or mercury, and the phases present are analyzed. The quenching method is well suited for those substances that react in a sluggish manner, for the quenching process itself does not usually cause any phase changes and the phases present at the higher temperatures from which the sample is quenched are still present after quenching.

An examination and analysis of phases present in samples quenched from different temperatures is made principally with the use of microscopic and X-ray diffraction techniques. As the sample of a given composition is quenched from successively higher temperatures, the temperature at which glass (liquid) first appears in the quenched sample is taken as the solidus temperature, and the temperature at which crystals no longer appear is the indicated liquidus temperature. The quenching method locates most precisely the liquidus temperatures or melting points in those systems that are slow to crystallize, such as borates, phosphates, and silicates.

Samples prepared for phase equilibrium studies are carefully weighed, ground in clean agate mortars, and mixed and ground together. The samples are heated to dispel water and carbon dioxide and to effect a small degree of sintering. The sample is then crushed and reground. With care taken to prevent the introduction of impurities, this process is repeated three or four times to approach as closely as possible a uniform, pure sample.

Since in many cases the quenching method does not "freeze in" the phases existing at high temperatures, a direct method of determining phases actually present at high temperatures is often used. The X-ray diffraction spectrometer, to which a high-temperature furnace has been adapted, permits direct analysis of crystalline phases present at these higher temperatures.

The light microscope with a hot stage has been adapted as an important technique for studying high-temperature reactions. This technique offers direct observation of melting and crystallization behavior and the opportunity to make a direct optical analysis of the crystalline phases.

It should be emphasized that normally no one technique of studying phase equilibria may be completely satisfactory; rather, several methods are used to study the phase equilibrium relations in a system, the final evaluation of phase equilibria relations being based on information derived from several techniques of investigation. For example, X-ray diffraction analysis, while providing qualitative data on the crystalline phases, may not be sufficiently precise in locating the liquidus temperature; the quenching method using the light microscope may indicate more accurately the presence or absence of crystals. The scanning electron microscope with an accessory hot stage has been shown to be another useful tool in studying phase relations.

Phase equilibria determination by the diffusion couple technique has not been very widely used in ceramic systems. Attainment of equilibrium in a diffusion couple of the end-members of a phase system can be possible if the couple is heated at sufficiently high temperature for interdiffusion to occur. Microscopic and quantitative microchemical analyses of the phases in the diffusion zone then permit the evaluation of the phase equilibria in the system. An application of these procedures is shown in Fig. 4.2, which shows the concentration profiles and the equilibria in the system SiO_2-Al_2O_3.

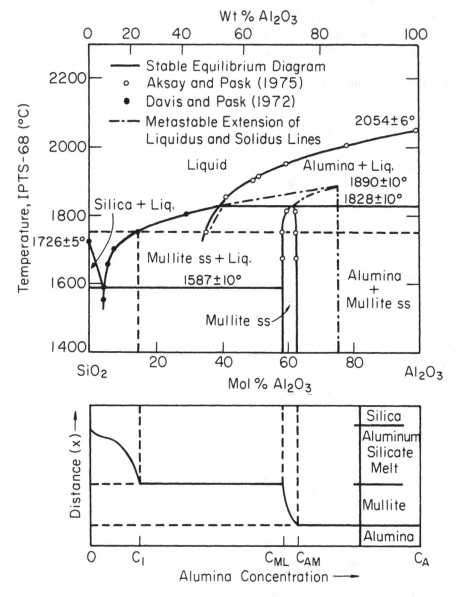

Fig. 4.2. SiO$_2$-Al$_2$O$_3$ stable equilibrium phase diagram (top). The relation between the stable phase diagram and the concentration profile of a semiinfinite SiO$_2$-Al$_2$O$_3$ diffusion couple is shown at temperature T below the melting point of mullite (bottom). Metastable extensions of the SiO$_2$-mullite system and the Al$_2$O$_3$ liquidus are superimposed on the stable SiO$_2$-Al$_2$O$_3$ diagram; C_A, 100% Al$_2$O$_3$; C_{ML}, concentration of Al$_2$O$_3$ in mullite at the Al$_2$O$_3$-mullite interface; C_I, concentration of Al$_2$O$_3$ in liquid saturated with mullite (after I. A. Aksay and J. A. Pask, *J. Am. Ceram. Soc.*, **58** [11–12] 507–12 (1975); and R. F. Davis and J. A. Pask, *ibid.*, **55** [10] 525–31 (1972)).

4.2. Thermodynamic Calculations and Estimations

A system is at equilibrium when its free energy is at a minimum. If we could calculate the free energy of all the possible phases of a system at a specified temperature as a function of composition, it would be a simple matter to select that phase or combination of phases which provides the lowest value of free energy for the system. By definition these would be the equilibrium phases for the system at that temperature. By repetition of these calculations for the number of temperatures, the phase boundaries of the system could be determined and the phase diagram could be constructed.

Because phase relations in ceramics are usually studied at constant pressure, the Gibbs free energy is used to determine the stable phases. The appropriate equation is

$$G = H - TS \tag{4.1}$$

where G = Gibbs free energy of the system under consideration, H = enthalpy, and T = absolute temperature (degrees Kelvin).

As mentioned earlier, the absolute values of these thermodynamic quantities are usually not known, but the changes in these quantities can be determined. By use of the expression $\Delta G = \Delta H - T\Delta S$ (at constant temperature), the change in free energy of the system can be related to changes in enthalpy and entropy. By convention, the free energies and enthalpies of the elements in their standard states* are assigned the value of zero at 298 K and 1 atm of pressure. The standard free energy of formation of a compound is the change in free energy accompanying the formation of the compound from its elements at 25°C and 1 atm of pressure. To determine the free energy change for the reaction at some other temperature, it is necessary to know the heat capacity, C_p, as a function of temperature for each of the reactants and products in order to calculate the changes in the enthalpy and entropy.

The change in the enthalpy with respect to temperature is given by

$$\left(\frac{\partial(\Delta H)}{\partial T} \right)_p = C_p(\text{products}) - C_p(\text{reactants}) = \Delta C_p \tag{4.2}$$

$$\Delta H^\circ_{T_2} - \Delta H^\circ_{T_1} = \int_{T_1}^{T_2} \Delta C_p dT \tag{4.3}$$

The corresponding change in the entropy term is given by

$$\Delta S^\circ_{T_2} - \Delta S^\circ_{T_1} = \int_{T_1}^{T_2} \frac{\Delta C_p}{T} dT \tag{4.4}$$

The calculation of the free energy change accompanying the formation of the compound can thus be calculated at any selected temperature. However, the standard free energies of formation are known for only a relatively small number of compounds, and the heat capacity data have been determined with accuracy over a wide temperature range for very few compounds. In practice, therefore, it is necessary to make a number of simplifying assumptions in order to perform the calculations. Several methods, which involve varying degrees of simplification, are described in the following pages.

Depression of the Freezing Point of a Solution

If the assumption is made that a mixture of two components, A and B, forms an ideal solution, the attractive forces between like and unlike atoms are equal. Raoult's law is obeyed and the partial vapor pressure of component A at temperature T is proportional to its concentration in the solution.

$$P_A = \frac{n_A}{n_A + n_B} P^\circ_A \tag{4.5}$$

where n_A = number of moles of A in solution, n_B = number of moles of B in solution, P°_A = vapor pressure of A in its pure state at temperature T.

If component A is assumed to be a pure solvent and B is a nonvolatile solute, the changes in vapor pressure with temperature for the pure solvent and for a dilute solution of B in A are as shown in Fig. 4.3.

The change in the vapor pressure of the solid A with changes in temperature is given by the Clausius-Clapeyron equation:

$$\frac{dP}{dT} = \frac{\Delta H_s}{T(V_v - V_s)} \tag{4.6}$$

where ΔH_s = enthalpy of sublimation of solid A, V_v = molar volume of vapor, and V_s = molar volume of solid.

*Standard states are designated by the superscript degree ($^\circ$), as H°, S°, G°.

53

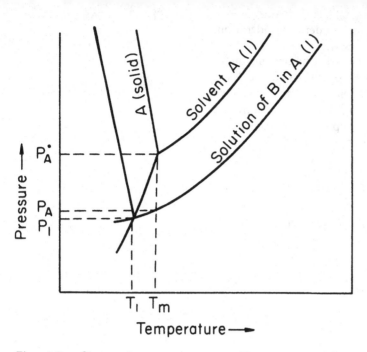

Fig. 4.3. Change in vapor pressure with temperature for a pure solvent and a dilute solution.

Owing to the enormous difference in molar volume between the solid and its vapor, the term V_s can be neglected. Further, if it is assumed that the vapor obeys the ideal gas laws and that enthalpy of vaporization is independent of temperature and concentration, the equation can be simplified as follows:

$$\frac{dP}{dT} = \frac{\Delta H_v}{TV_v}$$

and

$$V_v = \frac{RT}{P}$$

Therefore

$$\frac{dP}{dT} = \frac{\Delta H_v P}{RT^2} \tag{4.7}$$

Rearranging and integrating between limits

$$\int_{P_1}^{P_A^\circ} \frac{dP}{P} = \int_{T}^{T_M} \frac{\Delta H_v}{RT^2} dT$$

$$\ln \frac{P_A^\circ}{P_1} = \frac{\Delta H_v}{R} \left(\frac{1}{T} - \frac{1}{T_M} \right) \tag{4.8}$$

Assuming a similar relation for the solution of B in A

$$\ln \frac{P_A}{P_1} = \frac{\Delta H_s}{R} \left(\frac{1}{T} - \frac{1}{T_M} \right) \tag{4.9}$$

where P_A = the vapor pressure of A over the solution of B in A and ΔH_s = enthalpy of vaporization for the liquid A.

Subtracting Eq. (4.9) from Eq. (4.8)

$$\ln \frac{P_A^\circ}{P_1} - \ln \frac{P_A}{P_1} = \left(\frac{\Delta H_v}{R} - \frac{\Delta H_s}{R} \right) \left(\frac{1}{T} - \frac{1}{T_M} \right)$$

54

$$\ln \frac{P_A^\circ}{P_A} = \left(\frac{\Delta H_v - \Delta H_s}{R}\right)\left(\frac{1}{T} - \frac{1}{T_M}\right)$$

$\Delta H_v - \Delta H_s = \Delta H_f$ (from Hess' law)

where ΔH_f = enthalpy of fusion of A.

$$\ln \frac{P_A^\circ}{P_A} = \frac{\Delta H_f}{R}\left(\frac{1}{T} - \frac{1}{T_M}\right) \qquad (4.10)$$

From Eq. (4.5)

$$\frac{P_A^\circ}{P_A} = \frac{n_A + n_B}{n_A}$$

and

$$\ln \frac{P_A^\circ}{P_A} = \ln \frac{n_A + n_B}{n_A} = \frac{\Delta H_f}{R}\left(\frac{1}{T} - \frac{1}{T_M}\right)$$

but

$$\frac{n_A + n_B}{n_A} = \frac{1}{X_A}$$

where X_A = the mole fraction of A in the mixture of A and B.

Therefore

$$\ln \frac{1}{X_A} = \frac{\Delta H_f}{R}\left(\frac{1}{T} - \frac{1}{T_M}\right)$$

$$\ln X_A = -\frac{\Delta H_f}{R}\left(\frac{T_M - T}{T_M T}\right) \qquad (4.11)$$

$(T_M - T) = \Delta T$, the magnitude of the depression of the freezing point.

Equation (4.11) may be used to calculate the liquidus lines for a binary system. The results of such calculations for the system NaF-KF are given in Table 4.1. A comparison of the calculated liquidus lines with those determined by experiment is shown in Fig. 4.4. In this system the agreement is remarkably good, which is the exception rather than the rule for most ceramic systems.

The equation for the depression of the freezing point for dilute solutions was derived with the assumption that an ideal solution was formed, and therefore, the nature of the solvent and solute molecules was immaterial. If, however, the solvent or solute dissociates on melting to form a larger number

Table 4.1. Freezing Point Depressions Calculated from Eq. (4.11)

Temp (K)	X_A	ΔT
NaF (mp = 990°C; ΔH_f = 7780 cal/mol)		
1250	0.97	13
1200	0.85	63
1150	0.74	113
1100	0.63	163
1050	0.53	213
1000	0.45	263
KF (mp = 856°C; ΔH_f = 6750 cal/mol)		
1073	0.85	56
1023	0.73	106
973	0.63	156

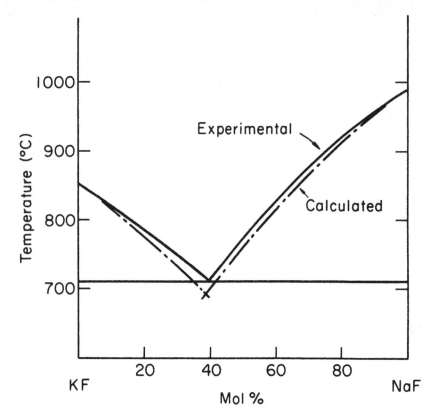

Fig. 4.4. Comparison of experimentally determined liquidus lines with those calculated from the freezing point depression equation.

of smaller particles or associates to form a fewer number of larger particles, the calculated liquidus line would be expected to deviate from the experimentally determined liquidus line. In Fig. 4.5 the liquidus lines were calculated from Eq. (4.11) by using enthalpy of fusion values calculated according to the method described in the last section of this chapter. If the intermediate compound AB is assumed to dissociate completely on melting, the total number of particles in the melt is increased and the effective mole fraction of solute atoms is decreased; hence, the depression of the freezing point of the

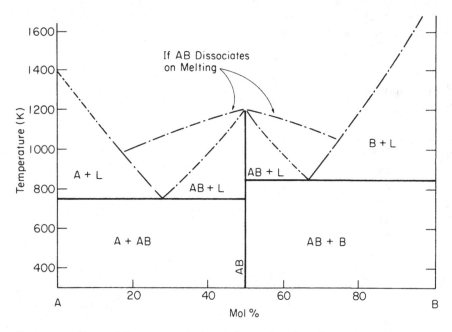

Fig. 4.5. Phase diagram calculated from the freezing point depression equation $\ln X_A = (-\Delta H_f/R)[(T_M - T)/(T_m T)]$.

solution is decreased as shown in Fig. 4.5. The broad maximum in the liquidus curves for the compound AB is indicative of dissociation of the compound on melting. The sharp maximum indicates that dissociation of the compound does not occur on melting.

In the system $PbO-B_2O_3$ (Fib. 3.29), the compound $PbO \cdot 2B_2O_3$ exhibits a very broad, flat maximum at the intersection of the liquidus curves, indicating dissociation on melting.

Calculation of Free Energy Differences Between Solid and Liquid Phases

In a binary system A-B, it is assumed that the components are completely miscible in the liquid state and completely immiscible in the solid state. The free energy of a mixture of solid A and solid B in which no solid solution occurs is shown in Fig. 4.6. With no reaction between the components, the free energy is simply that for a mechanical mixture of A and B.

If one assumes that the components are completely soluble in one another, the linear relation shown in Fig. 4.6 no longer holds because the free energy of the system is decreased by the free energy change which accompanies the solution reaction. If an ideal solution is formed, the enthalpy change for the reaction is zero and the entropy change is that due to mixing.

$$\Delta S = - R[X \ln X + (1 - X) \ln (1 - X)] \tag{4.12}$$

where ΔS = entropy of mixing, X = mole fraction of solute, and R = universal gas constant.

The free energy change for the reaction is given by

$$\Delta G = \Delta H - T \Delta S \text{ (at constant temperature)}$$

and since $\Delta H = 0$

$$\Delta G = -T(-R)[X \ln X + (1 - X) \ln (1 - X)]$$

$$\Delta G = RT [X \ln X + (1 - X) \ln (1 - X)] \tag{4.13}$$

Figure 4.7 shows a plot of the change in free energy for a mixture of A and B in which an ideal solution is formed.

The calculation of the changes in free energy accompanying the formulation of the liquid and solid phases in the system requires extensive thermodynamic data. However, it is possible to calculate *differences* in free energy between the liquid and solid phases of the same composition at selected temperatures by using a relation which is derived as follows:

$$G = H - TS \tag{4.14}$$

$$H = E + PV \tag{4.15}$$

$$G = E + PV - TS \tag{4.16}$$

where G = Gibbs free energy, H = enthalpy, S = entropy, T = absolute temperature, E = internal energy, P = pressure, and V = volume

$$dG = dE + PdV + VdP - TdS - SdT \tag{4.17}$$

For the reversible expansion of a perfect gas

$$dE = q_{rev} - w = q_{rev} - PdV \tag{4.18}$$

where q_{rev} = heat absorbed reversibly by the system and w = work done by the system.

Since

$$dS = \frac{q_{rev}}{T}$$

$$dE = TdS - PdV \tag{4.19}$$

Substituting Eq. (4.19) in Eq. (4.17)

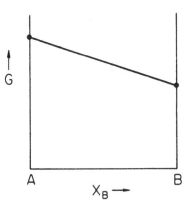

Fig. 4.6. Free energy of a mixture of A and B in which no solution occurs.

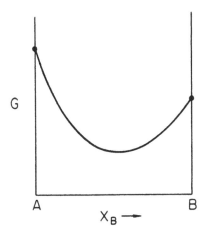

Fig. 4.7. Change in free energy for a mixture of A and B in which an ideal solution occurs.

$$dG = TdS - PdV + PdV + VdP - TdS - SdT$$

$$dG = VdP - SdT \tag{4.20}$$

At constant pressure

$$\left(\frac{\partial G}{\partial T} \right)_p = - S \tag{4.21}$$

The entropy change for the reaction solid→liquid at constant pressure and composition is given by

$$\left(S_l - S_s \right) = - \left[\left(\frac{\partial G_l}{\partial T} \right)_p - \left(-\frac{\partial G_s}{\partial T} \right)_p \right] \tag{4.22}$$

$$\Delta S = - \left(\frac{\partial (\Delta G)}{\partial T} \right)_p \tag{4.23}$$

where

$$\Delta G = (G_l - G_s)$$

At the melting point of the solid, equilibrium exists between solid and liquid and the change in free energy for the system is given by

$$\Delta G = \Delta H - T\Delta S = 0 \tag{4.24}$$

$$\Delta H_f = T\Delta S \tag{4.25}$$

$$\Delta S = \Delta H_f / T \tag{4.26}$$

where ΔH_f = the enthalpy of melting of the solid.

At constant pressure

$$\frac{d(\Delta G)}{dT} = - \Delta S \tag{4.27}$$

Substituting $\Delta H_f / T$ for ΔS in Eq. (4.27) yields

$$\frac{d(\Delta G)}{dT} = - \frac{\Delta H_f}{T}$$

Integrating between limits

$$\int_{\Delta G_T}^{0} d(\Delta G) = - \int_{T}^{T_M} \frac{\Delta H_f}{T} dT$$

$$- \Delta G = - \Delta H_f \ln \frac{T_M}{T}$$

$$- (G_l - G_s) = - \Delta H_f \ln \frac{T_M}{T} \tag{4.28}$$

$$G_s - G_l = - \Delta H_f \ln \frac{T_M}{T} \tag{4.29}$$

Equation (4.29) permits the calculation of the difference in free energy between the solid and liquid phases of a given substance at any temperature, providing the melting point of the substance and its enthalpy of fusion are known. The equation becomes less accurate when the temperature is farther from the melting point of the substance.

To illustrate this method, the phase diagram for the system BeO-UO₂ will be constructed from the calculated free energy data and will be compared with the published diagram.

In Fig. 4.8(A), pure UO_2 and pure BeO in their liquid states have been chosen as the points of reference for calculating the free energy differences between liquid and solid phases. The free energy change associated with the

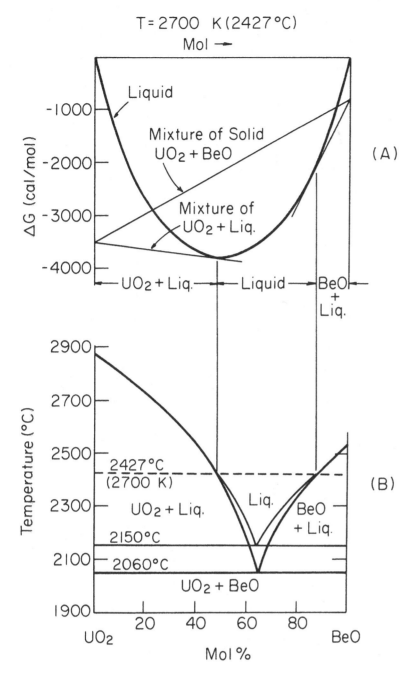

T= 2700 K (2427°C)

Mol →

Liquid

Mixture of Solid
UO_2 + BeO

(A)

Mixture of
UO_2 + Liq.

ΔG (cal/mol)

−1000

−2000

−3000

−4000

— UO_2 + Liq.— — Liquid — BeO —
+
Liq.

Temperature (°C)

2900

2700

2500

2427°C
(2700 K)

(B)

2300

UO_2 + Liq.

Liq.

BeO
+ Liq.

2150°C

2100

2060°C

UO_2 + BeO

1900

20   40   60   80

UO_2 BeO

Mol %

Fig. 4.8.   Determination of phase diagram from diagram of free
energy vs composition.

mixing of liquid UO_2 and liquid BeO at 2700 K was calculated from Eq.
(4.13); the results are plotted as the curve labeled "liquid" in Fig. 4.8(A).

The differences in free energy between liquid BeO and solid BeO and between the solid and liquid phases of UO_2 at 2700 K were calculated from Eq.
(4.29) by using the following values:

BeO, mp = 2830 K, ΔH_f = 17 000 cal/mol
UO_2, mp = 3150 K, ΔH_f = 22 900 cal/mol

At 2700 K, $G_s - G_l$ = −3530 cal/mol for UO_2 and −799 cal/mol for
BeO. These values are plotted on the graph of Fig. 4.8(A); the straight line
connecting these values represents $G_s - G_l$ for a mechanical mixture of solid
UO_2 and BeO.

The tangent drawn to the liquid curve at approximately 48% BeO (52%
UO_2) gives the composition of the liquid phase in equilibrium with UO_2. For
all compositions in the range 100–52% UO_2, a mixture of solid UO_2 and a melt

59

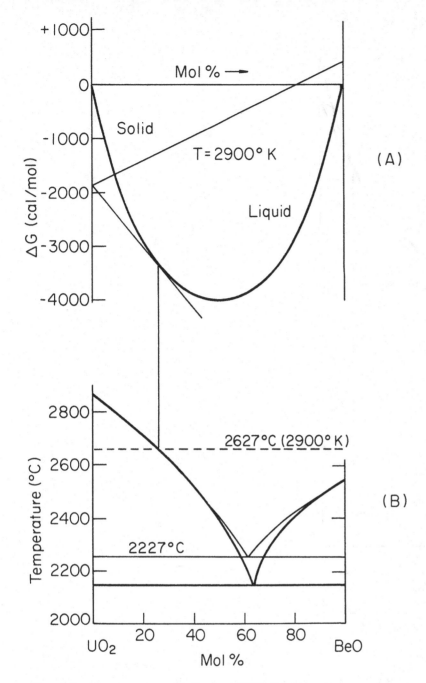

Fig. 4.9.   Free energy relations at 2900 K.

of 48% BeO/52% UO$_2$ will have the lowest free energy and will therefore be the stable phases at that temperature.

For compositions in the range 48–86% BeO, the liquid phase is the stable phase. At compositions greater than 86% BeO (the second point of tangency to the liquid curve in Fig. 4.8(A)), mixtures of solid BeO and a liquid composed of 86% BeO/14% UO$_2$ have the lowest free energy and are the stable phases. The points of tangency to the liquid curve in Fig. 4.8(A) have been extended into Fig. 4.8(B) to show the relation to the phase diagram.

By construction of free energy-composition diagrams at several temperatures, the compositions of the stable phases can be determined and the phase diagram can be subsequently plotted. Figures 4.9 and 4.10 show the data calculated for 2900 and 2500 K. The latter temperature is approximately that of the eutectic where UO$_2$/BeO, and a liquid of 65% BeO, 35% UO$_2$ coexist at equilibrium.

In Fig. 4.11 the experimentally determined diagram is shown with the

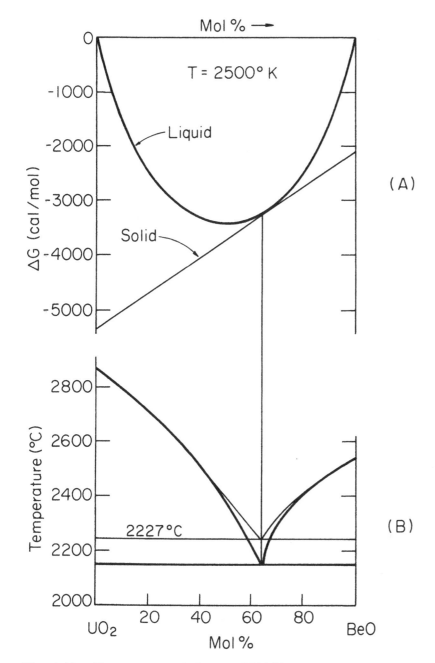

Fig. 4.10. Free energy relations at 2500 K.

calculated liquidus lines superimposed. The agreement is quite good, which suggests that the system approximates the ideal behavior assumed in the calculations.

Systems with Solid Solutions

The system NiO-MgO will be used to illustrate the application of calculation methods to a system in which the components are mutually soluble in both the liquid and solid states. The melting points and enthalpies of fusion for the components are as follows:

NiO, mp = 2257 K, $\Delta H_f = 12\ 100$ cal/mol
MgO, mp = 3073 K, $\Delta H_f = 18\ 500$ cal/mol

The free energy–composition curves are determined in a manner similar to that described in the preceding section. Figure 4.12 shows the free energy–composition diagram for 2773 K (2500°C). Equation (4.13) was used to calculate the free energy change associated with the mixing of liquid NiO and liquid MgO at the given temperature. Equation (4.29) was used to calculate the difference in free energy between the liquid and solid phases of each component of the system. A straight line connecting these two points represents

61

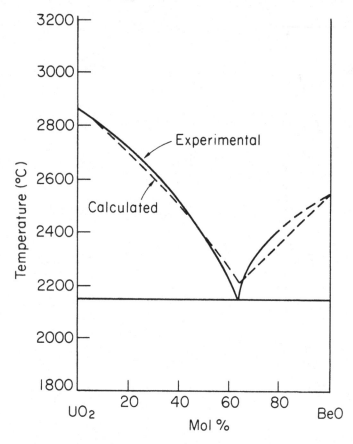

Fig. 4.11. Comparison of calculated and experimental liquidus curves.

the free energy difference between a mechanical mixture of solid MgO and solid NiO and their corresponding liquid-phase mixtures. The equation for this straight line (determined from the slope–intercept) is

$$\Delta G = (\Delta G_{MgO} - \Delta G_{NiO})X_{MgO} + \Delta G_{NiO} \tag{4.30}$$

where ΔG_{MgO} and ΔG_{NiO} are the values calculated from Eq. (4.29) and X_{MgO} is the mole fraction of MgO.

At 2500°C the equation for the system NiO-MgO was calculated to be

$$\Delta G = -4430X_{MgO} + 2500 \tag{4.31}$$

The curve which represents the free energy change associated with the solution of NiO and MgO in the solid state is determined by plotting the sum of the values of ΔG calculated from Eqs. (4.13) and (4.31). As seen in Fig. 4.12, a line drawn tangent to the curves for solid and liquid, respectively, indicates the composition of the stable phases. Similar diagrams constructed for temperatures of 2700 and 2200°C are shown in Fig. 4.13 and Fig. 4.14. The comparison between the experimentally determined diagram and that calculated in the manner just described is shown in Fig. 4.15.

In the foregoing examples, the assumption that ideal solutions were formed from the components of the system was found to be a fairly good approximation. Very often, however, our attempts to determine the equilibrium phases by such calculations yielded poor agreement with the experimental data because the solutions which are formed depart to a considerable extent from an ideal solution. If the departures from ideality are known and can thus be accounted for by activity coefficients, the calculations of free energy changes can be made more reliable.

In a qualitative way it is possible to account for some of the departures from ideality by analyzing the attractive forces in a solution of A and B atoms. Three kinds of bonds are considered:

62

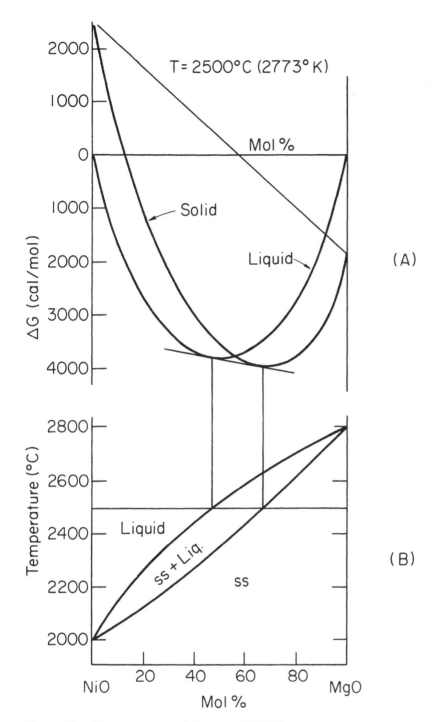

Fig. 4.12. Free energy relations at 2773 K.

A–A (bonds between pairs of A atoms)
B–B (bonds between pairs of B atoms)
A–B (bonds between A and B atoms)

Each bond type will have a characteristic enthalpy or bond energy associated with it. The enthalpy of a homogeneous solution of A and B atoms is given by

$$H = n_{AA}H_{AA} + n_{BB}H_{BB} + n_{AB}H_{AB} \tag{4.32}$$

where n_{AA}, n_{BB}, and n_{AB} represent the numbers of the respective bond types and H_{AA}, H_{BB}, and H_{AB} represent the associated enthalpies.

To determine the number of each kind of bond present, we assume some structural arrangement or lattice so that each atom has a fixed number of neighbors, either A or B.

63

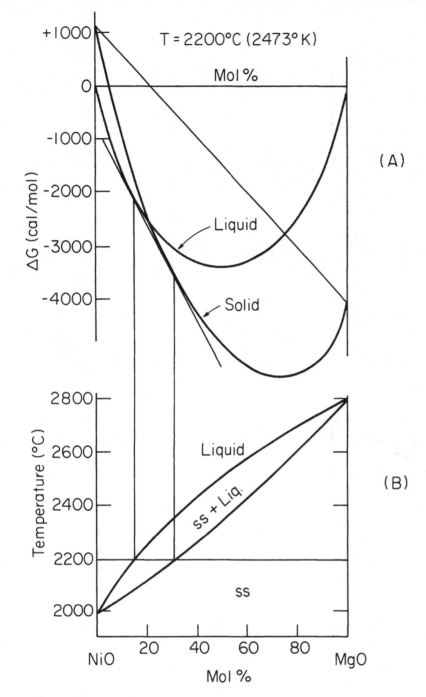

Fig. 4.13. Free energy relations at 2473 K.

The probability of finding an A atom on a particular lattice point is equal to its concentration, X_A, in the solution; similarly, that for B equals X_B. The probability of finding two A atoms next to each other is $X_A X_A = X_A^2$ and that for two B atoms is X_B^2. The probability of finding an A atom next to a B atom is $2(X_A X_B)$. The factor 2 is used because an A–B bond is indistinguishable from a B–A bond.

If the solution is considered to have a coordination number of z (each atom has z nearest neighbors), and there are N atoms in the solution, the total number of bonds will be $\frac{1}{2}zN$. The factor $\frac{1}{2}$ occurs because each bond is shared by two neighboring atoms.

The number of A–A bonds is given by $X^2(\frac{1}{2}zN)$, and the number of A–B type bonds is $2(X_A X_B)\frac{1}{2}zN$. The enthalpy of the solution is given by

$$H = \frac{1}{2}zNX_A^2 H_{AA} + \frac{1}{2}zNX_B^2 H_{BB} + 2(X_A X_B)\frac{1}{2}zNH_{AB}$$

64

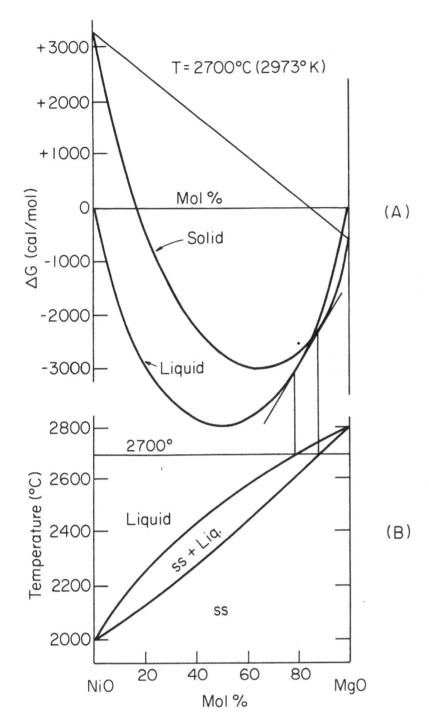

Fig. 4.14.   Free energy relations at 2973 K.

$$H = \frac{zN}{2}[X_A^2 H_{AA} + X_B^2 H_{BB} + 2(X_A X_B) H_{AB}] \qquad (4.33)$$

$$X_B = (1 - X_A)$$

$$H = \frac{zN}{2}[X_A^2 H_{AA} + (1 - X_A)^2 H_{BB} + 2X_A(1 - X_A)H_{AB}] \qquad (4.34)$$

Equation (4.34) can be reduced to the following expression:

$$H = \frac{zN}{2}\left[X_A H_{AA} + X_B H_{BB} + 2X_A X_B(H_{AB} - \frac{H_{AA} + H_{BB}}{2})\right] \qquad (4.35)$$

The first two terms on the right in Eq. (4.35) are equal to the enthalpy of a mechanical mixture of pure A and pure B. The reasoning is as follows: the en-

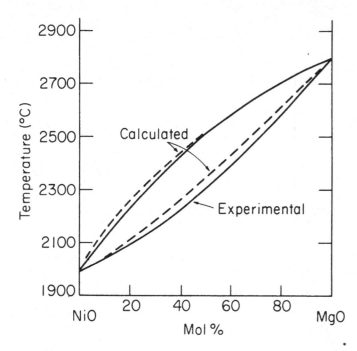

Fig. 4.15.   Comparison of calculated and experimentally determined phase diagram for system NiO-MgO.

thalpy of pure A ($X_A = 1$) is found from Eq. (4.33) to be $[(zN)/2](H_{AA})$. For pure B the enthalpy is $[(zN)/2](H_{BB})$. A mechanical mixture of A and B would have an enthalpy equal to

$$X_A H_A + X_B H_B = [(zN)/2][X_A H_{AA} + X_B H_{BB}] \qquad (4.36)$$

The last term in Eq. (4.35) represents the enthalpy of the interaction between A atoms and B atoms in the solution. It is a measure of the extent to which the solution departs from ideality and is called the enthalpy of mixing:

$$\Delta H_M = zN X_A X_B [H_{AB} - (H_{AA} + H_{BB})/2] \qquad (4.37)$$

Whether the enthalpy of mixing is positive or negative is dependent upon the type of interaction between A atoms and B atoms. If the attraction of A atoms for B atoms is greater than that of A for A or B for B, the enthalpy of mixing will be negative (binding energies are considered negative). If, on the other hand, the forces between A and B atoms are repulsive, the enthalpy of mixing will be positive.

The free energy change associated with the formation of a homogeneous solution of A and B atoms is given by

$$\Delta G_M = \Delta H_M - T \Delta S_M \qquad (4.38)$$

where ΔS_M is the entropy of mixing and equals $-Nk(X_A \ln X_A + X_B \ln X_B)$. These relations are graphically illustrated in Fig. 4.16.

When ΔH_M is negative (Fig. 4.16 (A)), dissimilar atoms attract more strongly than similar atoms and there exists a tendency toward the formation of an intermediate compound. When ΔH_M is positive (Fig. 4.16 (B)), there is a repulsion between unlike atoms and a tendency toward a clustering of like atoms or an unmixing of the solution to form two phases, either solid-solution phases or liquid-solution phases.

Estimation of the Enthalpy of Fusion

The enthalpies of fusion for substances of interest are frequently unknown, and it becomes necessary to make an estimate in order to proceed with the calculation of the stable phases. An empirical relation known as Richards' rule gives the entropy of fusion as

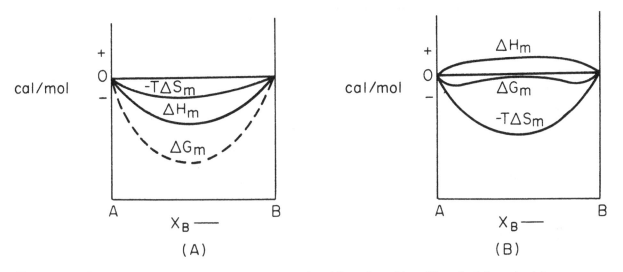

Fig. 4.16. Energy relations in system with negative (A) and positive (B) enthalpies of mixing.

$$\Delta S_f = \frac{\Delta H_f}{T_M} = 2 \tag{4.39}$$

when ΔH_f is in calories/g·atom.

For the elements of high atomic number, the relation holds fairly well (Au = 2.16, Ti = 2.25, Pb = 2.04, Th = 1.9, U = 2.28). For lower atomic number elements, some typical values are Na = 1.7, K = 1.64, Si = 6.6, B = 2.3.

To estimate the enthalpy of fusion for a compound, the melting process is considered to be comprised of two steps:

(1) $AB(s) \rightarrow AB(l)$
(2) $AB(l) \rightarrow A(l) + B(l)$

The first step corresponds to a hypothetical melting process in which solid AB melts to form liquid AB but retains the ordered structure characteristic of the solid. The second step represents the mixing of the elements of the compound to form a random solution. The entropy of mixing is therefore included in the calculation of the enthalpy of fusion.

For example: Calculate the enthalpy of fusion of UO_2.

(1) From Richards' rule: $\left(\dfrac{2\text{ eu}}{\text{mol}}\right)\left(\dfrac{3\text{ mol}}{UO_2}\right)$ = 6.0 eu

(2) Entropy of mixing $-3R[0.33 \ln (0.33) + 0.67 \ln (0.67)]$ = 3.78 eu

Total = 9.78 eu

$\Delta H_f = T_M \Delta S_f$
$\Delta H_f = 3150(9.78) = 30.8$ kcal/mol

The values determined by Grossman and Kaznoff from thermal analysis data ranged from 23.3 to 28.8 with an average value of 25.3 kcal/mol.

Thermodynamic Analysis of Liquidus Curves

Interesting thermodynamic approaches have been utilized to calculate activities of reacting components, metastable liquid immiscibility, and extensions of liquidus curves on phase diagrams in ceramic and glass forming systems. By use of assumptions based on the regular solution approximation, attempts have been made to calculate thermodynamic data in metal–slag systems, fused salts, and silicates. For example, Charles has outlined procedures for calculating activities in liquid SiO_2 solutions and for obtaining estimates of the liquid miscibility gaps on the relevant phase diagrams. Thermodynamic calculations using the liquidus curves in the important ceramic system SiO_2-Al_2O_3 led Risbud and Pask to propose two liquid miscibility gaps in the binary system SiO_2-Al_2O_3.

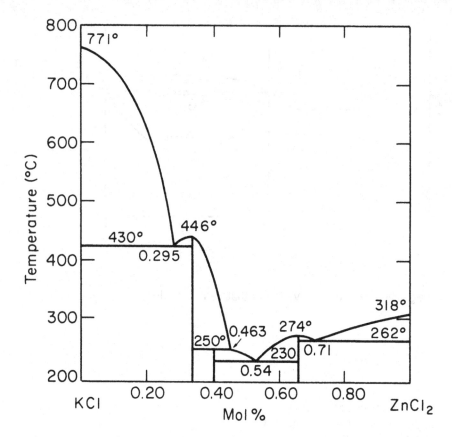

Fig. 4.17. Computer-calculated phase equilibrium for system KCl-ZnCl₂ (after P. L. Lin, A. D. Pelton, C. W. Bale, and W. T. Thomson; p. 47 in CALPHAD: Computer Coupling of Phase Diagrams and Thermochemistry, Vol 4. Pergamon, New York, 1980). (Reprinted by permission.)

The thermodynamic procedures for calculating data from the liquidus curves use the enthalpy of fusion, ΔH_f, which is assumed to stay constant with temperature, and the position of the liquidus curve on the phase diagram to obtain the activity of a component in a binary system through the equation:

$$\log_{10} a^l = \frac{\Delta H_f}{4.575}\left(\frac{1}{T_m} - \frac{1}{T_l}\right) = \log_{10} a^s \tag{4.40}$$

where a^l = activity of the liquid in the liquidus composition referred to pure liquid standard state, a^s = activity of solid at the liquidus temperature referred to pure solid standard state, ΔH_f = enthalpy of fusion, T_m = melting temperature (K) of the pure component, and T_L = liquidus temperature (K).

The data calculated along the liquidus at various temperatures and compositions can then be extrapolated to other temperatures by assuming that at a fixed composition the partial molar enthalpy of solution, $\Delta \bar{H}$, stays independent of temperature. Thus

$$RT_l \log_{10} \gamma_l = RT \log_{10} \gamma_T \tag{4.41}$$

where T_l and γ_l = liquidus temperature and activity coefficient at the liquidus, respectively, T and γ_T = temperature and activity coefficient at the chosen temperature T, R = gas constant = 1.98 cal/mol·deg (K). The activity a and the activity coefficient γ are related by the equation $a = \gamma x$, where x = mole fraction.

The calculated isothermal activity coefficients for one component can then be used in the Gibbs–Duhem equation to obtain data for the other component in the binary system:

$$X_1 d \log \gamma_1 + X_2 d \log \gamma_2 = 0 \tag{4.42}$$

where X_1 and X_2 = mole fractions of components 1 and 2 and γ_1 and γ_2 = activity coefficients of components 1 and 2, respectively. The calculated activities are

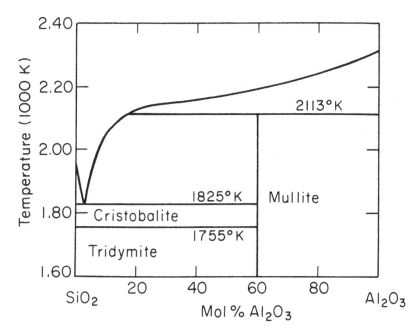

Fig. 4.18. Computer-calculated phase equilibrium for system SiO_2-Al_2O_3 (after P. Dorner, L. J. Gauckler, H. Kreig, H. L. Lukas, G. Petzow, and J. Weiss; CALPHAD: Computer Coupling of Phase Diagrams and Thermochemistry, **3**, 241 (1979)). (Reprinted by permission.)

where a_1 and a_2 = activities of components 1 and 2. The variation of free energy of mixing with composition is used to locate the width of the liquid miscibility gap at each temperature, therefore allowing calculation of this particular form of metastable phase equilibria between two supercooled liquids (glasses).

Computer-Assisted Phase Equilibria

As computers rapidly develop into "user friendly" machines, the trend toward computer calculations of phase equilibria continues to be on the rise. While such calculations are not likely to entirely eliminate the need for experimental determination of phase diagrams, they certainly promise to make the task of assembling phase equilibria information much less cumbersome and tedious. Indeed, the birth of a new technical journal[†] devoted entirely to phase diagram computer calculations testifies to the emerging interest in this approach.

The computer calculations of phase diagrams rely heavily on expressing the thermodynamic properties of solutions in the form of suitable analytical equations. Calculations are often performed on an interactive basis by calling a computer program via a terminal station. By feeding appropriate thermodynamic information, a calculated phase diagram is generated on the terminal screen. A special advantage of this interactive approach is that it permits a quick assessment of the influence of changing thermodynamic parameters on computed phase equilibrium curves.

Two examples of computer-calculated phase equilibria are shown in Figs. 4.17 and 4.18. Figure 4.17 shows the computer-calculated KCl-$ZnCl_2$ binary system utilizing an interactive computer program described by Lin, Pelton, Bale, and Thomson. The program used in this work is available for general use via the TELENET/TYMENET telecommunications network linked to a computer center based in Montreal, Canada. Using computer calculations based on the Newton–Raphson interaction method, Dormer, Gauckler, Kreig, Lukas, Petzow, and Weiss computed phase equilibria in several ceramic systems. Figure 4.18 is an example of the calculated SiO_2-Al_2O_3 phase equilibria from their work. This calculated diagram compares reasonably well with the experimentally determined equilibria shown in Fig. 4.2.

†CALPHAD: Computer Coupling of Phase Diagrams and Thermochemistry. Pergamon, New York.

Problems for Chapter 4

4.1. *System V_2O_5-NiO (Fig. 3.39).*
A crucible containing a melt of composition 45 mol% V_2O_5 is allowed to cool in air from 1000°C to 500°C. Construct a graph to show how the temperature would be expected to vary as a function of time during the cooling of this melt.

4.2. *Explain how the slope of the liquidus curve near the indifferent point of a congruently melting compound is influenced by the degree of dissociation of the compound when it melts.*

4.3. *Construct the phase diagram for the binary system CaO-MgO by calculating the difference in free energy between the liquid and solid phases at various temperatures. Assume the system to be a simple eutectic system. The melting point for CaO is 2630°C and its enthalpy of fusion is 12 000 cal/mol. For MgO, the melting point is 2825°C and the enthalpy of fusion is 18 500 cal/mol.*

4.4. *Estimate the enthalpies of fusion for $Li_2O \cdot B_2O_3$ and for $Na_2O \cdot B_2O_3$. Assume melting points of 850°C and 970°C, respectively.*

4.5. *Determine the slopes of the liquidus lines in the system $Li_2O \cdot B_2O_3$-$Na_2O \cdot B_2O_3$ from calculations of the freezing point depression. Assume a simple eutectic system. Compare your results with published data.*

4.6. *Estimate the enthalpies of fusion for $KNbO_3$ and $KTaO_3$. Assume melting points of 1064°C and 1372°C, respectively.*

4.7. *Determine the slopes of the liquidus and solidus lines for the system $KNbO_3$-$KTaO_3$. Assume complete solid solution between the end-members.*

Bibliography and Supplementary Reading

R. J. Charles, *J. Am. Ceram. Soc.*, **50** [12] 631 (1967).

P. Dorner, L. J. Gauckler, H. Kreig, H. I. Lukas, G. Petzow, and J. Weiss. *CALPHAD: Comput. Coupling Phase Diagrams Thermochem.*, **3**, 211 (1979).

W. Eitel; Thermochemical Methods in Silicate Investigation. Rutgers University Press, New Brunswick, NJ, 1952.

M. N. Grossman and A. I. Kaznoff, *J. Am. Ceram. Soc.*, **51** [1] 59–60 (1968).

L. Kaufman, *CALPHAD: Comput. Coupling Phase Diagrams. Thermochem.*, **3**, 275 (1979).

L. Kaufman and H. Bernstein; Computer Calculation of Phase Diagrams. Academic Press, New York, 1970.

W. J. Knapp, *J. Am. Ceram. Soc.*, **36** [2], 43 (1953).

P. L. Lin, A. D. Pelton, C. W. Bale, and W. T. Thomson, *CALPHAD: Comput. Coupling Phase Diagrams Thermochem.*, **4**, 47 (1980).

J. B. MacChesney and P. E. Rosenberg; p. 114–65 in Phase Diagrams, Vol. 6–1. Edited by A. M. Alper. Academic Press, New York, 1970.

J. A. Pask and I. A. Aksay, Transport Phenomena in Ceramics. Edited by A. R. Cooper, Jr. and A. H. Heuer, Plenum Press, New York, 1972.

S. H. Risbud and J. A. Pask, *J. Am. Ceram. Soc.*, **60** [9–10] 418 (1977).

R. Roy and O. F. Tuttle; Physics and Chemistry of the Earth. Edited by L. A. Ahrens, K. Rankama, and S. K. Rancorn, McGraw-Hill, New York, 1956.

J. H. Welch, *J. Sci. Instrum.*, **31**, 458 (1954).

F. E. Wright and G. A. Rankin, *Am. J. Sci.*, **39**, 1 (1915).

5.1. Hypothetical Binary Systems

One of the methods which is useful as an aid to becoming familiar with binary phase diagrams is that of constructing hypothetical systems from statements of the relations between phases. Following is an example:

Construct the binary phase diagram which represents the following conditions in the system A-B:

1. Component A melts at 1850°C.
2. Component B melts at 1700°C.
3. Compound A_4B melts incongruently at 1500°C.
4. Compound AB melts congruently at 1600°C.
5. Compound AB_3 is stable between 1200 and 1350°C.
6. A eutectic is formed between A_4B and AB at 1400°C.
7. A eutectic is formed between AB and B at 1450°C.

The solution is given in Fig. 5.1.

Many of the binary systems involve partial solid solution between two compounds, as shown in Fig. 5.2, where A is partially soluble in B to form the solid solution β, and B is partially soluble in A to form the solid solution α.

From the geometry of the free energy vs composition curves, Rhines has shown that the solidus line *ab* should be constructed such that its metastable extension *bc* would project into the two-phase region $\alpha + \beta$. Similarly, the subsolidus line *fb* should be constructed such that its metastable extension *bd* would extend into the two-phase region $\alpha + L$. An incorrect construction is shown in Fig. 5.3, where the solidus and solvus lines are drawn such that their metastable extensions would project into the single-phase region.

The argument is as follows: Consider the free energy-composition

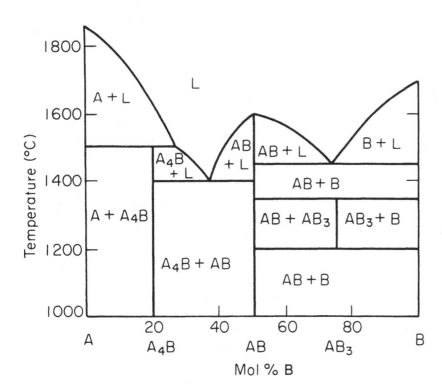

Fig. 5.1. Solution to problem.

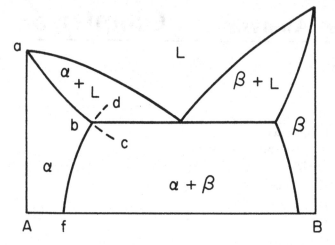

Fig. 5.2. Metastable extensions of solidus and solvus lines should project into two-phase regions.

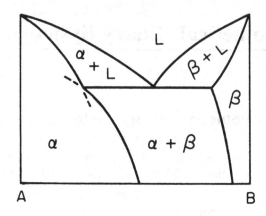

Fig. 5.3. Incorrect construction.

curves of Fig. 5.4 at temperature T_1 for the α and γ phases, respectively (lower diagram). Note that for compositions between g and h, a mixture of α and γ provides the lowest free energy, as shown by the tangent to the two curves. The proportion of α and γ for any composition between g and h is determined from the lever relationship. The points g and h are the limits of the miscibility gap on the phase diagram (middle diagram of Fig. 5.4) at temperature T_1.

At temperature T_2 (upper diagram), which is above the eutectic temperature, there is a region of composition extending from c to d in which the liquid phase has the lowest free energy. For compositions between a and c, a mixture of α and liquid of composition c has the lowest free energy. Between d and f, a mixture of γ and liquid of composition d has the lowest free energy.

The tangent points b and e in the upper diagram represent the compositions of the solid phases which would be in equilibrium if there were no intervening liquid phase of lower free energy. It is apparent from the diagram that for geometric reasons, point a must always lie to the left of point b and point f must always lie to the right of point e when a liquid phase of lower free energy is present. Thus, the solidus and subsolidus lines must be drawn so as to project into the two-phase regions.

5.2. Phase Analysis Diagrams*

It is often desirable to analyze the reactions which occur and the phases which are present during the melting of a given crystalline composition under equilibrium conditions. Such information is useful in estimating the firing range of a particular ceramic composition or in determining the safe operating range for a refractory material.

Most ceramic processes for producing a finished product begin with powdered materials that are subsequently formed into a shape by any of a number of methods such as dry pressing, slip casting, or extruding. Firing or heat treatment is used to consolidate or densify this shape in order to give it strength and other desirable properties. The changes which occur in the microstructure during the firing process are shown in Fig. 5.5.

Densification is accomplished by solid-state sintering or by sintering with the aid of a liquid phase. In the latter case, the liquid phase which is formed during firing aids in drawing the crystalline particles together and enhances the atomic diffusion involved in the solution and precipitation reactions occurring at the firing temperatures. The amount of liquid which is sufficient to permit densification without causing the ware to slump is dependent upon the fluidity of the melt. In general, the volume of liquid at the firing

---

*Other names for these diagrams include "column melting diagrams," "melting behavior charts," and "phase composition vs temperature diagrams."

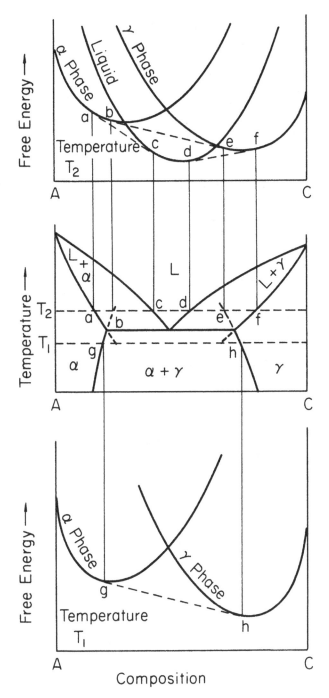

Fig. 5.4. Free energy-composition curves at two temperatures for system A-C.

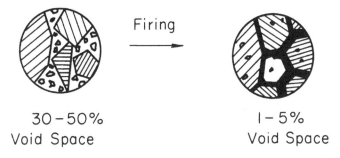

30 – 50%
Void Space

Firing

1 – 5%
Void Space

Fig. 5.5. Densification during firing.

temperature ranges from 20 to 50%. During cooling, some crystallization will occur from the melt and some of the melt will supercool to form a glassy or vitreous bond between particles.

In the isoplethal studies described in previous chapters, we determined the crystalline phases which occurred during the cooling of the melt under equilibrium conditions. It should be recognized that the path of crystallization of a given liquid which solidifies to certain crystalline phases is identical with the path of melting of these crystalline phases to the same given liquid. Thus, we can use the data obtained from the isoplethal study to plot a phase analysis diagram which shows the percentage of phases as a function of temperature. In Fig. 5.6, phase analysis diagrams have been drawn for three compositions in the system Al_2O_3-SiO_2. The diagram for the 55% Al_2O_3 melt represents a crystalline composition of approximately 10% SiO_2 and 90% mullite. If this mixture were heated, about 10% liquid would form at 1600°C. The amount of liquid would increase very slowly, as shown by the steep curve, until approximately 25% liquid was present at 1800°C. This very gradual increase in the amount of liquid as the temperature is increased is indicative of a long firing range.

The mixture of approximately 70% mullite and 30% Al_2O_3 (75% Al_2O_3 melt) would form no liquid until heated to 1840°C, at which temperature almost 80% of the sample would melt to form a liquid. A refractory of this composition would be stable below 1840°C, but its structural properties would be lost above that temperature.

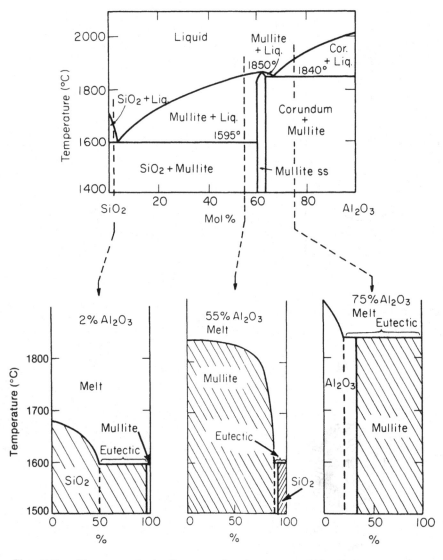

Fig. 5.6. Phase analysis diagrams for the system $SiO_2 \cdot Al_2O_3$.

Problems for Chapter 5

5.1. Construct the hypothetical binary system L-G which represents the following conditions (label all areas):

    a. Component L melts at 1600°C.

    b. Component G melts at 1700°C.

    c. Compound LG melts congruently at 1500°C.

    d. Compound LG_9 melts incongruently at 1450°C.

    e. A maximum of 20% LG is soluble in L forming the solid-solution phase γ.

    f. A maximum of 10% L is soluble in LG forming the solid-solution phase β.

    g. There is a eutectic between γ and β at 1300°C.

    h. Two liquids are in equilibrium above 1450°C between 55 and 70% of G.

    i. A eutectic exists between LG and LG_9 at 1300°C.

5.2. Construct the hypothetical binary system S-T which represents the following conditions (label all areas):

    a. Component S melts at 1800°C.

    b. Component T melts at 1550°C.

    c. Compound S_2T_3 melts congruently at 1450°C.

    d. Compound S_2T melts incongruently at 1400°C.

    e. A maximum of 25% of T is soluble in S_2T_3 forming the solid-solution phase α.

    f. A eutectic exists between α(ss) and T at 1300°C.

    g. $β$-S_2T transforms to $α$-S_2T when heated above 1150°C.

    h. A eutectic exists between $α$-S_2T and S_2T_3 at 1300°C.

5.3. Construct phase analysis diagrams for the following compositions:

    a. System V_2O_5-NiO (Fig. 3.39): (1) 45 mol% NiO, 55 mol% V_2O_5, (2) 60 mol% NiO, 40 mol% V_2O_5.

    b. System CaO-MgO (Fig. 3.22): 20 wt% CaO, 80 wt% MgO.

    c. System CaO-SiO_2 (Fig. 3.37): 40 wt% CaO, 60 wt% SiO_2.

Bibliography and Supplementary Reading

W. D. Kingery, H. K. Bowen, and D. R. Uhlmann; Introduction to Ceramics, 2d ed. Wiley, New York, 1976.

E. M. Levin, C. R. Robbins, and H. F. McMurdie, Phase Diagrams for Ceramists, 1964. Edited by M. K. Reser. The American Ceramic Society, Columbus, Ohio; Fig. 314.

A. Prince; Alloy Phase Equilibria. Elsevier, New York, 1966.

F. H. Rhines; Phase Diagrams in Metallurgy: Their Development and Applications. McGraw-Hill, New York, 1956.

Systems containing three components are called ternary systems. The phase rule becomes $F = 3 - P + 2$, and an invariant point is a position where five phases coexist. If pressure is constant, the phase rule for the system is $F = 4 - P$, and an invariant point involves equilibrium between four phases. The maximum number of solid and liquid phases which can coexist at equilibrium is thus four for a condensed ternary system.

The ternary system can be represented by a three-dimensional figure such as shown schematically in Fig. 6.1. The base consists of an equilateral triangle, the apices of which represent the composition of the three components. The temperature axis is perpendicular to this base. The top surface describes the topography of the liquidus which is a curved surface made up of hills and valleys. The sides of this solid represent the limiting binary systems.

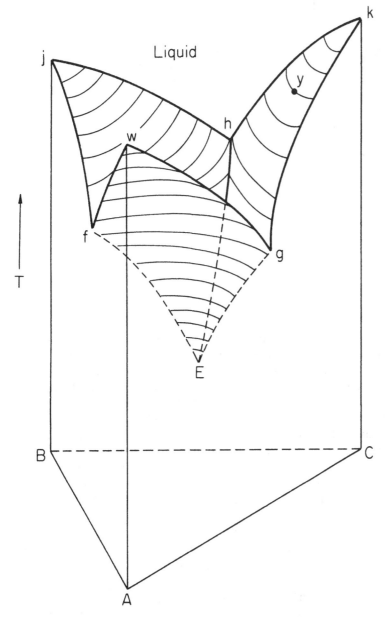

Fig. 6.1. Diagram of a ternary system containing one ternary eutectic.

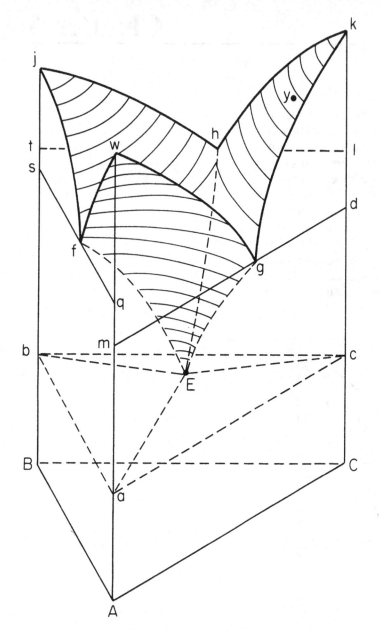

Fig. 6.2. Space figure for ternary system.

Fig. 6.3. Space of primary crystallization of A.

In Fig. 6.1 the limiting binary systems, A-B, B-C, C-A, each contain one binary eutectic, shown by the points *f, h,* and *g,* respectively.

The liquidus surface is the locus of all temperature-composition points representing the maximum solubility of a solid phase in a liquid phase. At temperatures above the liquidus surface, the equilibrium phase is liquid. Points on the liquidus surface represent equilibrium between a liquid and a solid phase. For example, point *y* located on the liquidus surface *khEg* represents equilibrium between a liquid and crystals of C. The intersections of the liquidus surfaces form the boundary lines *hE, fE,* and *gE*. Points on the boundary lines represent equilibrium between two crystalline phases and a liquid. Along the boundary line *hE*, crystals of B and C are in equilibrium with a liquid.

The boundary lines intersect at point *E*, which is the ternary eutectic. Crystals of A, B, and C are in equilibrium with a liquid at this point. Because there are four phases coexisting, the point *E* is an invariant point. For this system, the eutectic point represents the lowest melting mixture of crystals of A, B, and C. No liquid exists at temperatures below that of point *E*.

In Fig. 6.2 the solidus lines for the limiting binary systems of the ternary have been drawn. In addition, the horizontal plane *abc* which contains the eutectic point *E* has been constructed; lines have been drawn in this plane

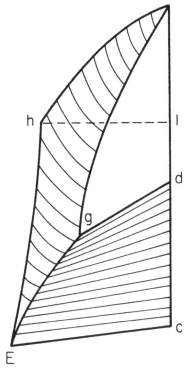

Fig. 6.4. Space of primary crystallization of C.

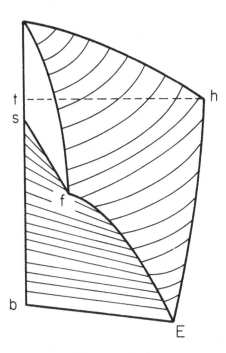

Fig. 6.5. Space of primary crystallization of B.

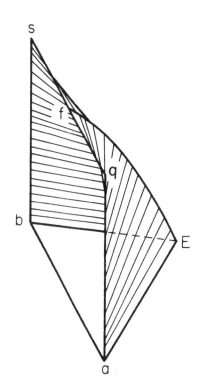

Fig. 6.6. Space of crystallization of A + B.

from E to the points a, b, c. Figure 6.2 now contains all of the external lines necessary to define the eight "volumes" or "spaces" of which the figure is composed, as follows:

1. The *liquid space* is the region above the liquidus surface and is composed of liquid only.

2. The *three spaces of primary crystallization*, A + liquid, B + liquid, and C + liquid, are bounded on their upper surfaces by the liquidus surface. Their lower surfaces are generated in the following manner: Consider the primary crystallization space of A + liquid. Place a slender rod, such as a pencil, in a horizontal position with one end contacting the vertical line Aw at point m and the other end contacting the boundary line gf at point g. If the left end of the rod is permitted to slide from m to a while being maintained horizontally and in contact with the boundary line gE as the right end slides to point E, the surface thus generated is part of the lower bounding surface of the primary crystallization space of A. The remainder of the lower surface is generated by sliding the rod horizontally from q to a while maintaining contact with boundary line fE. The space thus generated is shown in Fig. 6.3. All composition-temperature points within this space represent equilibrium between crystals of A and liquids whose compositions lie on the liquidus surface $wfEg$. The lower surface of the primary crystallization spaces of B + liquid and C + liquid are generated in a manner similar to that described for A + liquid and are shown in Figs. 6.4 and 6.5.

3. The *three spaces of binary crystallization*, A + B + liquid, A + C + liquid, and B + C + liquid, are bounded by the lower surfaces of the spaces of primary crystallization and the horizontal plane abc which contains the eutectic point E. These three "cavelike" spaces are shown in Figs. 6.6, 6.7, and 6.8. All composition-temperature points within these spaces represent equilibrium between two crystalline phases and a liquid, the composition of which will lie on the appropriate boundary line (either gE, fE, or hE). For example, in the binary crystallization space of A + C + liquid, all liquid compositions will lie on the boundary line gE. The vertical outer surfaces of these spaces are in the binary systems at temperatures below that of the binary solidus and hence contain only two crystalline phases.

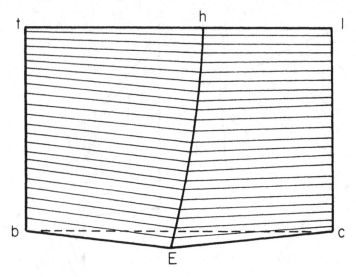

Fig. 6.7. Space of crystallization of B + C.

4. The *space containing crystalline phases only* is that space below the plane *abc* and is composed of crystals of A, B, and C. Figure 6.9 shows these spaces in relation to the ternary figure representing the system.

6.1. Method of Determining Composition

The equilateral triangle which forms the base of the ternary figure permits the representation of all possible combinations of the three components. There are two frequently used methods for determining the composition of a point within the triangle.

1. The lengths of the perpendiculars drawn from the sides of the composition triangle to the point are proportional to the quantities of the crystals in the sample. For example, in Fig. 6.10 the composition X is composed of 40% C, 20% A, and 40% B.

2. Lines are drawn through the composition point and parallel to the sides of the triangle. The intersection of these lines with any side gives the proportions of the crystalline phases represented by the point. For example, in Fig. 6.11 the point Y is composed of 10% A, 30% B, and 60% C. This method is also applicable to nonequilateral triangles.

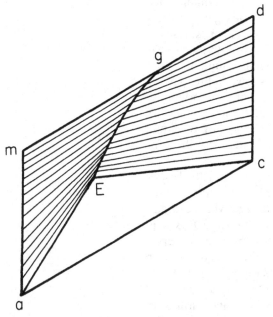

Fig. 6.8. Space of simultaneous crystallization of A + C.

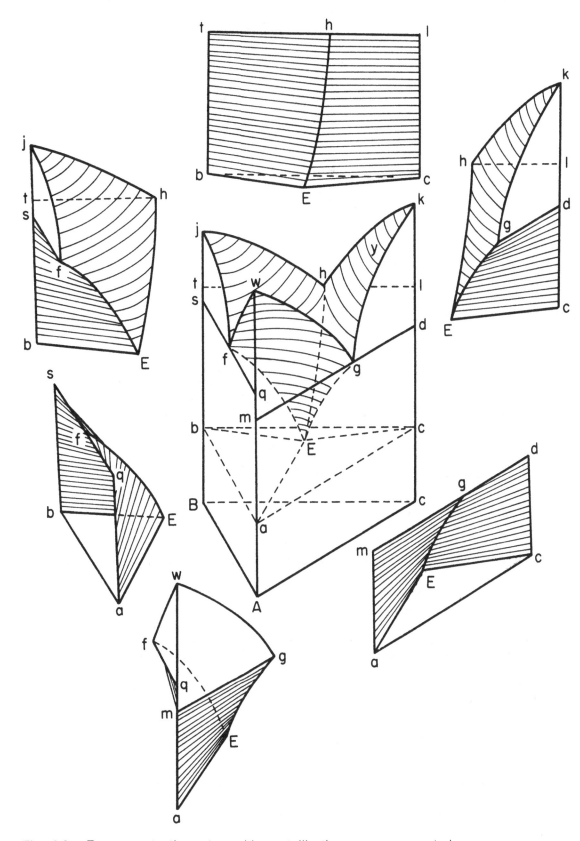

Fig. 6.9. Ternary eutectic system with crystallization spaces separated.

6.2. Isoplethal Studies in Ternary Systems (Example 6.1)

The changes which occur during the cooling of a melt of a given composition are usually worked out through the use of a plane projection of the liquidus surface. However, the rules of construction which are used on the plane projection have been derived from the three-dimensional solid representing the ternary system. Therefore, it is desirable to understand the

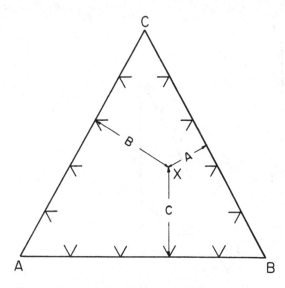

Fig. 6.10. Method for determining the composition of a point within the ternary system A-B-C.

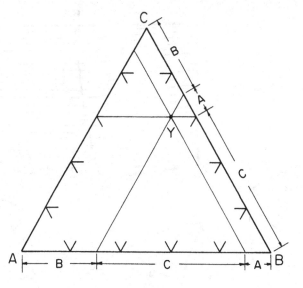

Fig. 6.11. Alternate method for determining composition of a point within a composition triangle.

correlation between the three-dimensional figure and its plane projection. In Fig. 6.12 an isopleth (XX') has been constructed for the composition X. The plane projection of the liquidus surface is shown at the base of the three-dimensional figure. The temperature scale on the three-dimensional figure is shown by a series of isothermal lines on the liquidus surface of the primary crystallization space of component C. These isotherms have been projected into the plane projection at the base of the figure.

The intersection of the isopleth with the liquidus surface occurs at temperature T_3 and is shown by the point labeled 3. A tie line, $T_3 - 3$, has been constructed to join the conjugate phases (melt of composition 3 and crystals of C). The corresponding projection onto the plane projection is denoted by the point X' and the line CX'. As the melt of composition X cools to temperature T_4, crystals of C precipitate from the melt and the concentration of C in the melt decreases. The conjugate phases are again determined by drawing the tie line (in a direction away from the crystallization component, C) from T_4 through the isopleth at 4 and on to its intersection with the liquidus surface at 4'. The position of 4' within the composition triangle $A - B - C$ gives the composition of the melt at temperature T_4. Further cooling of the sample of overall composition X to temperature T_5 results in additional precipitation of crystals of C and a corresponding change in the melt composition to that given by point 5'.

At temperature T_6, the composition of the melt corresponds to that given by point 6' which is located on the boundary line HE. The isopleth is in contact with one of the lower surfaces of the primary crystallization space of C (point 6) (also see Fig. 6.4). (An examination of the vertical section shown in Fig. 6.13 may help to clarify this relation.) Further cooling now places the composition into the space of binary crystallization, C + B + liquid. At temperature T_7 the composition of the melt is given by the point at 7', and the crystalline portion of the sample is composed of B and C.

As the sample is cooled to the eutectic temperature, the melt composition moves along the boundary line from 7' to E. Simultaneously, additional quantities of C and B precipitate from the melt. At a temperature just above the eutectic temperature, crystals of C and B are in equilibrium with a melt of composition E. As the temperature is decreased slightly below that of E, the melt crystallizes in eutectic proportions. That is, crystals of A, B, and C precipitate simultaneously until all of the melt has been solidified.

Quantitative calculations of the amounts of the phases present at each temperature are made with the use of the lever rule and are most easily

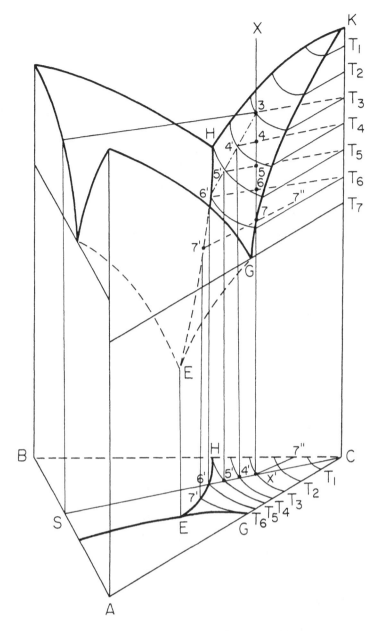

Fig. 6.12.   Cooling path for melt of composition X.

followed on the plane projection of the liquidus surface. For convenience this plane projection has been reproduced in Fig. 6.14 and will be used in the following analysis. At temperature T_3 the sample is composed of melt plus an infinitesimal quantity of crystals of C. When the sample is cooled to temperature T_4, additional quantities of C are precipitated. The quantities are determined by drawing a tie line from C through point X and terminating on the T_4 isotherm at point 4'. The precentages of melt and crystal are determined from the lever rule, as follows:

$$\frac{CX}{C4'}(100) = \% \text{ melt of composition } 4'$$

$$\frac{X4'}{C4'}(100) = \% \text{ crystals of C}$$

At temperature T_5

$$\frac{CX}{C5'}(100) = \% \text{ melt of composition } 5'$$

$$\frac{X5'}{C5'}(100) = \% \text{ crystals of C}$$

83

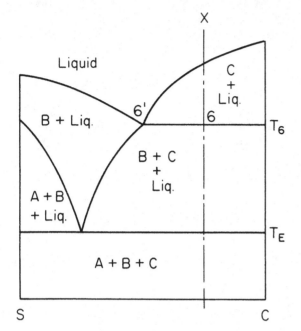

Fig. 6.13. Vertical section through system shown in Fig. 6.12 (not a true binary).

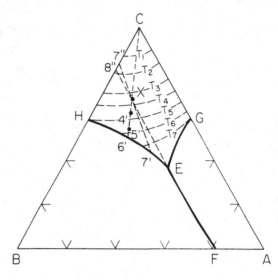

Fig. 6.14. Plane projection of liquidus surface of system shown in Fig. 6.12.

At temperature T_6

$$\frac{CX}{C6'}(100) = \%\text{ melt of composition } 6'$$

$$\frac{X6'}{C6'}(100) = \%\text{ crystals of C}$$

At T_6 an infinitesimal quantity of crystals of B appears, and on cooling from T_6 to T_7, the quantity of B crystals increases. The new tie line is drawn from $7'$ through X and on to its intersection with the line BC and point $7''$. This intersection divides the line BC into two lengths which are used to determine the relative proportions of B and C crystals in the sample at temperature T_7. At temperature T_7

$$\frac{7''X}{7''7'}(100) = \%\text{ melt of composition } 7'$$

$$\frac{X7'}{7''7'}(100) = \%\text{ crystals}\begin{cases} \dfrac{7''B}{CB}(100) = \%\text{C} \\[2mm] \dfrac{7''C}{CB}(100) = \%\text{B} \end{cases}$$

When the mixture is cooled to a temperature just above the eutectic temperature, additional precipitation of B and C occurs and the quantity of melt decreases.

At T_E^+ (temperature slightly above the eutectic temperature)

$$\frac{8''X}{8''E}(100) = \%\text{ melt of composition } E$$

$$\frac{XE}{8''E}(100) = \%\text{ crystals}\begin{cases} \dfrac{8''B}{CB}(100) = \%\text{C} \\[2mm] \dfrac{8''C}{CB}(100) = \%\text{B} \end{cases}$$

At a temperature slightly below the eutectic temperature, the melt solidifies in eutectic proportions. The proportions of A, B, and C in the eutectic mixture may be determined by either of the methods described earlier.

At T_E^- (temperature slightly below the eutectic temperature)

$$\frac{8''X}{8''E}(100) = \%\text{ eutectic}\begin{cases} 44\%\ \text{A} \\ 36\%\ \text{C} \\ 20\%\ \text{B} \end{cases}$$

84

$$\frac{XE}{8''E}(100) = \% \text{ crystals} \begin{cases} \dfrac{8''B}{CB}(100) = \%C \\[2mm] \dfrac{C8''}{CB}(100) = \%B \end{cases}$$

The cooling path for any melt (composed of A, B, and C) in the system will terminate at the eutectic point E. The nature of the crystals which precipitate as the melt cools will depend upon the location of the sample composition within the ternary diagram. The areas $BHEF$, $AFEG$, and $CHEG$ of Fig. 6.14 are called the *primary phase fields** of B, A, and C, respectively. For a melt whose composition falls within the primary phase field of B, the first crystal to appear on cooling will be that of B. Conversely, B will be the last crystal to melt on heating. Along the boundary lines HE, GE, and FE, two crystalline phases are in equilibrium with a melt. For example, along the boundary line FE crystals of A and B are in equilibrium with a melt whose composition varies along the boundary line.

6.3. Alkemade Lines

A straight line connecting the points representing the compositions of the crystals in the ternary system is called a *binary join*. A join connecting the compositions of the primary crystals of two areas having a common boundary line is called an *Alkemade line*. The use of Alkemade lines simplifies the analysis and interpretation of ternary phase relations. The following example should help to clarify the distinction between joins and Alkemade lines.

In the system shown in Fig. 6.15, there is an intermediate binary compound BC which has a primary phase field labeled *bc*. The line *A-BC* is a join

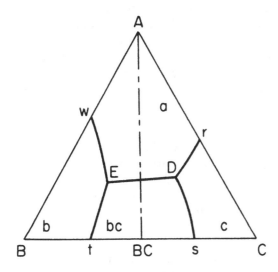

Fig. 6.15. Join B-C is not an Alkemade line.

connecting the compositions of the primary crystals of two areas having a common boundary line, *ED*, and therefore, the line *A-BC* is an Alkemade line. Lines *AC*, *AB*, *B-BC*, and *BC-C* are also joins connecting the compositions of the primary crystals of areas having a common boundary line and are, therefore, also Alkemade lines. Line *BC*, although a join, does not fulfill the condition of the common boundary line because areas *c* and *b* are not adjacent; consequently, *BC* is not an Alkemade line.

The Alkemade theorem states that the intersection of a boundary line (or boundary line extended) with its corresponding Alkemade line (or Alkemade line extended) represents a temperature maximum on that boundary line and a temperature minimum on the Alkemade line. The Alkemade theorem permits one to analyze with respect to temperature the direction of slope of boundary lines and to determine the general trend in the shape of the liquidus

*Sometimes called "primary fields" or "fields of primary crystallization."

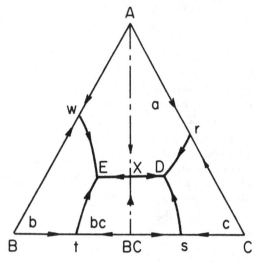

Fig. 6.16. Application of the Alkemade theorem to determine slopes of boundary lines.

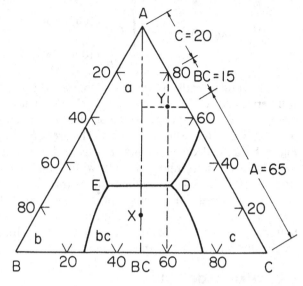

Fig. 6.17. Solidification of melt Y produces a mixture of 20% C crystals, 15% BC crystals, and 65% A crystals.

surface. In Fig. 6.16 the slopes of the boundary lines for the system A-B-C have been determined by applying the Alkemade theorem. Note the intersection of the Alkemade line A-BC with its corresponding boundary line ED at point X. This intersection represents a temperature maximum on the boundary line, and therefore, the boundary line slopes from X toward E and from X toward D. As indicated by the slope of the Alkemade line A-BC, the liquidus surface of the primary phase field of A slopes toward X, as does the primary phase field of BC. Three boundary lines slope to a common junction at E and at D. These points are eutectic points. Which of the two eutectic points is lower in temperature cannot be determined from the Alkemade theorem.

6.4. Composition Triangles

Alkemade lines divide a ternary diagram into *composition triangles.*[†] In the diagram of Fig. 6.15, triangles *A-BC-B* and *A-BC-C* are composition triangles. A composition triangle is always made up of three Alkemade lines. It should be noted that Alkemade lines never cross one another.

When cooled to solidification under equilibrium conditions, any melt whose composition falls within a composition triangle will form a combination of the crystals indicated at the apices of the composition triangle. Thus, the final crystals and their proportions can be determined for any melt cooled to solidification under equilibrium conditions. For example, in Fig. 6.17 composition Y is located within the composition triangle *A-BC-C*. When melt Y solidifies, the sample will be composed of the crystals represented at the apices of this composition triangle. The relative proportions of these three final crystalline phases are determined (on the composition triangle) by drawing lines through point Y and parallel to the sides of the composition triangle, as shown in Fig. 6.17. The intersection of these lines with any side of the composition triangle indicates the proportions of the crystals present. For point Y in Fig. 6.17, the composition of the final crystals is $A = 65\%$, $BC = 15\%$, and $C = 20\%$. A portion of the crystals will, of course, be contained in the eutectic structure. The composition of the eutectic is determined from the position of point D in the composition triangle and is found to be $A = 30\%$, $BC = 48\%$, and $C = 22\%$.

If a melt has the composition of the eutectic corresponding to the composition triangle in which it occurs, the melt will solidify in eutectic proportions at the eutectic temperature.

[†]Sometimes called "compatibility triangles."

86

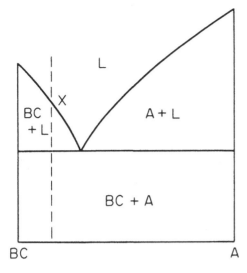

Fig. 6.18. Vertical section through the ternary system of Fig. 6.17.

If a melt does not occur in a composition triangle, it will be on an Alkemade line and will finally solidify as the crystals represented by the extremities of the Alkemade line and in a proportion indicated by its position on the line. The cooling path of the melt will follow along the Alkemade line to its intersection with the corresponding boundary line. For example, the melt of composition X in Fig. 6.17 is located on the Alkemade line A-BC. On cooling, the melt will change composition along the line A-BC toward the line ED as crystals of BC precipitate from the melt. At a temperature corresponding to the intersection of the line A-BC with the boundary line ED, the melt will solidify. The proportions of A and BC are given by the location of the point X on the line A-BC and are $A = 16\%$, $BC = 84\%$. Because the **Alkemade line A-BC intersects its corresponding boundary line, ED, the system is divided into two ternary systems, A-BC-C and A-BC-B. Thus, the line A-BC represents a true binary system and may be treated as such.** A vertical section along A-BC of the three-dimensional figure represented by Fig. 6.17 is shown in Fig. 6.18 and will help to illustrate the cooling path of the melt given in the example above.

If the Alkemade line does not intersect its corresponding boundary line, the crystalline phases represented by the ends of the Alkemade line do not represent a true binary system. This situation is discussed in Section 6.9.

6.5. Isothermal Sections

The construction of isothermal sections of the three-dimensional ternary figure is helpful in gaining a better understanding of the relation of the Alkemade lines on the liquidus surface to the crystallization spaces of the three-dimensional figure. In Fig. 6.19 the isotherms have been drawn on the ternary system of Fig. 6.17. An isothermal section at 700°C is shown in Fig. 6.20. The primary crystallization spaces of A, B, and C and the liquid space have all been intersected by the 700°C plane. The lines radiating from the composition points in the primary phase fields are tie lines and are usually drawn in to identify the two-phase regions more readily.

Figure 6.21 is an isothermal section at 600°C. The primary crystallization space of BC is intersected at this temperature, as is the space of binary crystallization, A + C + L.

At 400°C, which is below the temperature corresponding to the intersection of the Alkemade line A-BC with its boundary line, the isothermal section (Fig. 6.22) shows the separation of the system into two separate ternary systems. Except for the space composed of crystals only, all of the spaces in

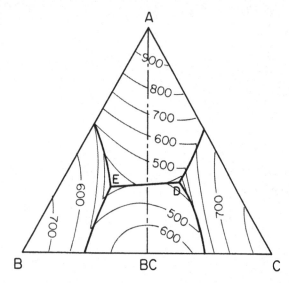

Fig. 6.19. System A-B-C with isotherms added.

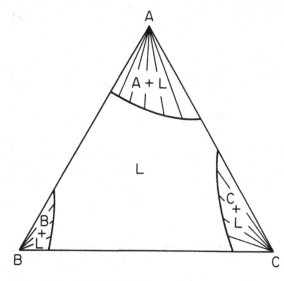

Fig. 6.20. Isothermal section at 700°C.

the three-dimensional figure appear in this section. The Alkemade line is seen to be a vertical plane, the upper limit of which is the temperature corresponding to that of the intersection with the boundary line *ED*. Figure 6.23 shows the section at 300°C, which is below the temperature of either eutectic point.

6.6. System with a Binary Compound Melting Incongruently

Figure 6.24 illustrates a system similar to that shown in Fig. 6.19, but only one ternary eutectic forms because the Alkemade line *A-BC* does not intersect its boundary line. The boundary line *DE* slopes away from its Alkemade line *A-BC* from the point *D* toward *E*. A melt, such as that indicated by point *X*, occurring in the composition triangle *A-BC-C* will, on

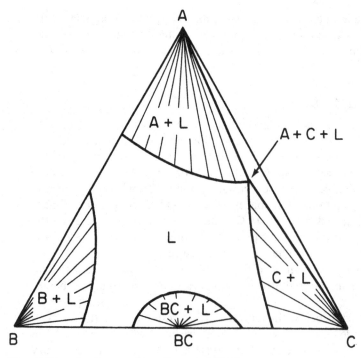

Fig. 6.21. Isothermal section at 600°C.

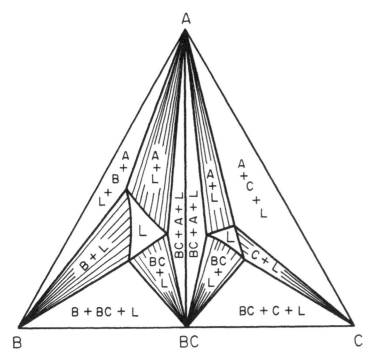

Fig. 6.22.   Isothermal section at 400°C.

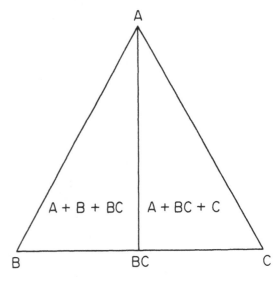

Fig. 6.23.   Isothermal section at 300°C.

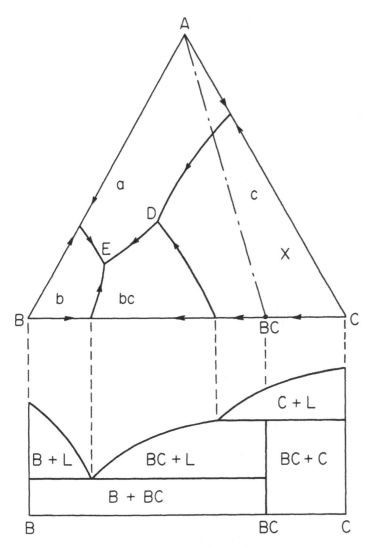

Fig. 6.24.   Ternary system with a binary compound melting incongruently. Binary system for side B-C is shown below.

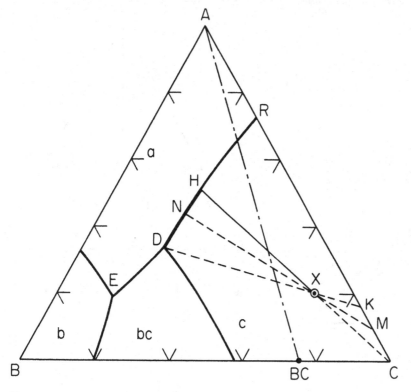

Fig. 6.25. Cooling path of melt of composition X is from X to H to D.

cooling to solidification, be made up of crystals of A, BC, and C. The only point at which these three crystals are in equilibrium is at point D, which is outside the composition triangle A-BC-C. Point D is a peritectic point (not a eutectic point); it is the point to which the melt will cool and also represents the temperature of final solidification of the melt. The final crystals, however, when solidification of the melt of composition X is complete, will be composed of A, BC, and C in proportions determined by the position of point X within the composition triangle A-BC-C. The following isoplethal study will serve to illustrate the sequence of reactions:

Isoplethal Study (Example 6.2)

1. *Final Crystals.* Composition X ($A = 20\%$, $B = 10\%$, $C = 70\%$) is located within the composition triangle A-BC-C of Fig. 6.25. When solidification is complete, the sample will be composed of crystals of A, BC, and C. The proportions can be determined by drawing lines through point X parallel to the sides of the composition triangle A-BC-C and noting the intersections along the line AC (or either of the other two sides of the composition triangle). The composition of the final crystals is $A = 20\%$, $BC = 40\%$, and $C = 40\%$. These figures are recorded on the calculation sheet of Table 6.1.

2. *Cooling Path of Melt.* A. The melt will cool to the point at which crystals of A, BC, and C are in equilibrium. This is seen to be at the peritectic point D. Composition X is located within the primary phase field of C; therefore, the first crystal to precipitate on cooling the melt is that of C. When C precipitates from the melt, the concentration of C in the melt is decreased and the composition of the melt changes along a line through point X and directly away from point C, as shown on the diagram. This tie line intersects the boundary line DR at point H. The proportions of crystals and melt at this temperature are determined by applying the lever rule. The melt composition in terms of components A, B, and C is determined by the position of point H within the triangle A-B-C. These calculations are shown in Table 6.1.

90

Table 6.1. Calculation sheet for ternary systems (See Fig. 6.25)

$$\text{Melt} \begin{cases} A=20 \\ B=10 \\ C=70 \end{cases} \qquad \text{Final Crystals} \begin{cases} A=20 \\ BC=40 \\ C=40 \end{cases}$$

Temp.	Proportions and Composition of Phases		Analysis A	B	C
1 (Point X)	$\frac{40}{40}$ Melt=100%	$\begin{cases} A=20 \\ B=10 \\ C=70 \end{cases}$	20	10	70
	†ε Crystals = ε% C				
2 (Point H)	40 Melt=39%	$\begin{cases} A=51 \\ B=27 \\ C=22 \end{cases}$	20	10	9
	$\frac{62}{102}$ Crystals=61% C ε% A				$\frac{61}{70}$
3 (Point N)	26 Melt=30%	$\begin{cases} A=44 \\ B=34 \\ C=22 \end{cases}$	13	10	7
	$\frac{61}{87}$ Crystals=70%	$\begin{cases} A=\dfrac{10}{100}=10 \\[4pt] C=\dfrac{90}{100}=90 \end{cases}$	$\frac{7}{20}$		$\frac{63}{70}$
4 (Slightly above point D)	19 Melt=23%	$\begin{cases} A=32 \\ B=45 \\ C=23 \end{cases}$	7	10	6
	$\frac{64}{83}$ Crystals=77%	$\begin{cases} A=\dfrac{17}{100}=17 \\[4pt] C=\dfrac{83}{100}=83 \end{cases}$	$\frac{13}{20}$		$\frac{64}{70}$
	Peritectic Reaction Occurs: $C(s)+A,B,C(\ell)\rightleftarrows BC(s)+A(s)$				
5 (Slightly below point D)	Yielding final crystals	$\begin{cases} A=20 \\ BC=40 \\ C=40 \end{cases}$	20	10 30	$\frac{40}{70}$

†ε = infinitesimal amount of.

B. Along the boundary line RD, crystals of A and C are in equilibrium with the melt. When the sample is cooled further, the composition of the melt changes from H toward D along the boundary line HD. Simultaneously, crystals of A and C precipitate from the melt. At a temperature corresponding to point N, the tie line MN can be drawn (the selection of point N was arbitrary; there are an infinite number of tie lines that can be drawn between the temperatures represented by points H and D). The proportions of solid and melt are determined by applying the lever rule. The solid portion of the sample, however, is now composed of two kinds of crystals, C and A. The proportions of C and A are determined by applying the lever rule along the Alkemade line AC. The intersection at M separates the two lever arms, AM and MC. AM represents the proportion of C, and MC represents the proportion of A crystals in the solid portion of the sample. (See Table 6.1.)

C. At a temperature slightly above the peritectic temperature at D, the tie line DK can be drawn. The proportions of melt and crystals of A and C are calculated as shown in Table 6.1. Note that the proportion of crystals of A increased as the sample cooled to the lower temperature.

D. At a temperature just slightly below the peritectic temperature, the peritectic[‡] reaction occurs. During this reaction the melt reacts with one of the crystals which is present and forms two different crystalline substances. In this case, the melt reacts with a portion of the C crystals to form crystals of BC and A. The reaction may be written as follows:

$$C(s) + A,B,C(l) \leftrightharpoons BC(s) + A(s)$$

Since four phases are involved, we know from a consideration of the phase rule that the reaction must be isothermal. When the reaction is complete, the sample will contain crystals of A, BC, and C in the proportions determined by the position of the point X within its composition triangle, as was discussed earlier.

3. *Cooling Path of Solid.* As the molten sample cools to complete solidification, the compositional changes of the solid portion can be followed by noting the positions of the intersections of successive tie lines on the ap-

Table 6.2. Calculation Sheet for Ternary Systems (See Fig. 6.27)

Melt $\begin{cases} A=25 \\ B=40 \\ C=35 \end{cases}$ Final Crystals $\begin{cases} A=11.5 \\ AB=53.5 \\ C=35 \end{cases}$

	Temp.	Proportions and Composition of Phases	A	B	C
Y	79	Melt=100% $\begin{cases} A=25 \\ B=40 \\ C=35 \end{cases}$	25	40	35
		ε Crystals (B)			
G	79	Melt=80% $\begin{cases} A=31 \\ B=25 \\ C=44 \end{cases}$	25	20	35
	$\frac{19}{98}$	Crystals=20% (B, C) $\begin{cases} B=100 \\ C=\varepsilon \end{cases}$		$\frac{20}{40}$	
P^+	40	Melt=68% $\begin{cases} A=37 \\ B=26 \\ C=37 \end{cases}$	25	18	25
	$\frac{19}{59}$	Crystals=32% (B, C) $\begin{cases} \frac{68}{100}B=68 \\ \frac{32}{100}C=32 \end{cases}$		$\frac{22}{40}$	$\frac{10}{35}$
		Peritectic reaction occurs: B(s) + A,B,C(ℓ)⇌AB(s) + C(s)			
P^-	13	Melt=41% $\begin{cases} A=37 \\ B=26 \\ C=37 \end{cases}$	15	10.5	15
	$\frac{19}{32}$	Crystals=59% (AB, C) $\begin{cases} \frac{89}{135}AB=66\ (39) \begin{cases} A=25 \\ B=75 \end{cases} \\ \frac{46}{135}C=34\ (20) \end{cases}$	$\frac{10}{25}$	$\frac{29.5}{40.0}$	$\frac{20}{35}$
F	12	Melt=27% $\begin{cases} A=49 \\ B=21 \\ C=30 \end{cases}$	13.5	6	8
	$\frac{32}{44}$	Crystals=73% (AB, C) $\begin{cases} \frac{85}{135}AB=63\ (46) \begin{cases} A=25 \\ B=75 \end{cases} \\ \frac{50}{135}C=37\ (27) \end{cases}$	$\frac{11.5}{25.0}$	$\frac{34}{40}$	$\frac{27}{35}$

(Table 6.2 continued next page)

[‡]Some authors consider this reaction to be intermediate between a eutectic and a peritectic reaction and reserve the term peritectic for reactions of the type $L + A + B \rightleftarrows C$.

				Analysis		

Table 6.2 (continued).

| Melt $\begin{cases} A=25 \\ B=40 \\ C=35 \end{cases}$ | | | Final Crystals $\begin{cases} A=11.5 \\ BC=53.5 \\ C=35 \end{cases}$ | | | |

Temp.	Proportions and Composition of Phases			Analysis		
				A	B	C
E^+ 12	Melt=21%	$\begin{cases} A=59 \\ B=17 \\ C=24 \end{cases}$		12.5	4	5
$\dfrac{45}{57}$	Crystals=79% (C, AB)	$\begin{cases} \dfrac{83}{135}AB=61.5\ (48.5) \\ \dfrac{52}{135}C=38.5\ (30.5) \end{cases}$	$\begin{cases} A=25 \\ B=75 \end{cases}$	$\dfrac{12.5}{25}$	$\dfrac{36}{40}$	$\dfrac{30}{35}$

Melt solidifies to form eutectic structure: $A,B,C(\ell) \rightleftarrows A(s) + AB(s) + C(s)$

| E^- 12 | Eutectic=21% | $\begin{cases} A=52 \\ AB=24\ (5) \\ C=24 \end{cases}$ | $\begin{cases} A=25 \\ B=75 \end{cases}$ | $\begin{matrix}11 \\ 1.5\end{matrix}$ | 4 | 5 |
| $\dfrac{45}{57}$ | Crystals=79% (C, AB) | $\begin{cases} \dfrac{83}{135}AB=61.5\ (48.5) \\ \dfrac{52}{135}C=38.5\ \ (30.5) \end{cases}$ | $\begin{cases} A=25 \\ B=75 \end{cases}$ | $\dfrac{12.5}{25.0}$ | $\dfrac{36}{40}$ | $\dfrac{30}{35}$ |

propriate Alkemade line. For example, the solid portion of sample X followed the path from C to M to K. During the peritectic reaction the solid portion changed from K to X. These changes are summarized in the phase analysis diagram of Fig. 6.26. Note on the diagram that when the sample is cooled from H to D, the composition of the solid portion of the sample increased from 61 to 77%, but that most of this increase was due to the precipitation of crystals of A; the concentration of C remained nearly constant during this period. The proportion of crystals which are precipitating at any given temperature can be determined by constructing a line tangent to the cooling path of the melt at that temperature and extending to the appropriate Alkemade line. For example, in Fig. 6.25 a line tangent to the boundary line RD at point N and intersecting at a point near R on the Alkemade line AC indicates the ratio of A and C crystals which are precipitating at temperature N. It can be estimated that the crystals which precipitated when the sample cooled from H to N were composed of about 75% A and 25% C, although the overall composition of the solid portion of the sample changed from 100% C to 90% C and 10% A.

6.7. Peritectic and Eutectic Reactions during Cooling

Before final solidification of the melt Y in Fig. 6.27, both a peritectic and a eutectic reaction will occur. The isoplethal study for melt Y is as follows:

Isoplethal Study (Example 6.3)

1. Final Crystals. Composition Y falls within the composition triangle A-C-AB and when solidified will be composed of crystals of A, C, and AB in the proportions determined by its position within the triangle. (See Table 6.2.)

2. Cooling Path of Melt. The melt will cool to the position at which crystals of A, C, and AB are in equilibrium (point E). Composition Y is located within the primary phase field of B; therefore, B will be the first crystal to precipitate on cooling. The composition of the melt will change along the line YG extending from point B through point Y to the point G

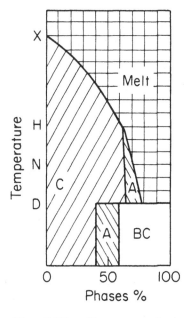

Fig. 6.26. Phase analysis diagram for composition X of Fig. 6.25.

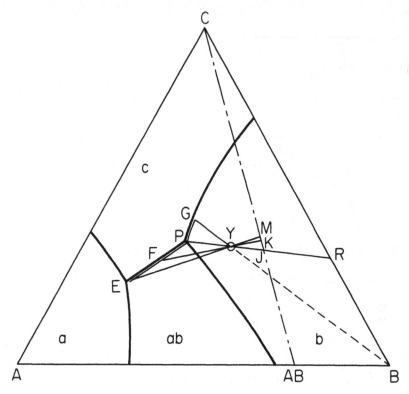

Fig. 6.27. Cooling path for a melt of composition Y. Both peritectic and eutectic reactions occur during cooling.

located on the boundary line between the primary phase fields of B and C.

Crystals of B and C will coprecipitate as the melt composition changes along the boundary line from G to P. At a temperature just slightly below that of point P, the peritectic reaction occurs; the melt reacts with one of the crystals present to form two different crystalline species. In this case, since C and AB are the crystalline species in equilibrium with melt along line PE, we know that all of the B crystals must react with the melt to form crystals of AB and some additional C. The reaction may be written as follows:

$$B(s) + A,B,C(l) \leftrightarrows AB(s) + C(s)$$

After the peritectic reaction is complete, further cooling will cause the melt to change along the boundary line from P to E. Since C and AB are the two crystals now in equilibrium with the melt, the tie line from point P terminates at point J, which is on the Alkemade line C-AB. The proportions of AB and C which are in equilibrium are determined by the position of this intersection. (See Table 6.2.)

At temperature F, the appropriate tie line for use in calculations is line FK. At a temperature just slightly above the eutectic temperature corresponding to point E, the appropriate tie line is EM. Crystals of C and AB are in equilibrium with a melt of composition E. At this temperature, 21% of the sample is melt. (See Table 6.2.)

At a temperature slightly below the eutectic temperature, the eutectic reaction occurs and the melt of composition E solidifies in eutectic proportions: A = 52%, AB = 24%, and C = 24%.

3. Cooling Path of Solid. The solid portion of the sample was composed of B crystals as the melt cooled from Y to G. The solid composition changed from point B to point R along the Alkemade line CB as the melt cooled from G to P. During the peritectic reaction, the composition of the solid changed from point R to point J on the Alkemade line C-AB. As the melt cooled from P to E, the solid changed in composition from J to M. During the eutectic reaction, the overall composition of the solid changed from point M to point Y. These changes are summarized in the phase analysis diagram of Fig. 6.28.

94

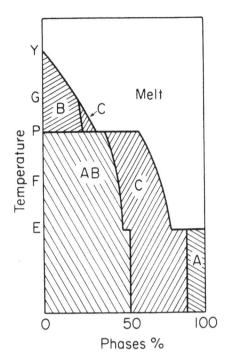

Fig. 6.28. Phase analysis diagram for composition Y of Fig. 6.27.

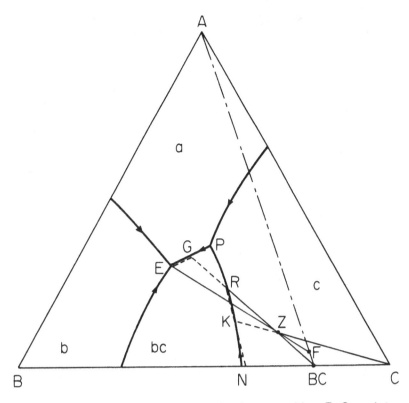

Fig. 6.29. Cooling path for a melt of composition Z. Complete resorption of primary crystalline phase C occurs during cooling.

6.8. Resorption during Cooling

During cooling of a sample composition Z (Fig. 6.29), there will occur a precipitation of C crystals which will subsequently be resorbed as cooling progresses. The analysis is as follows:

Isoplethal Study (Example 6.4)

1. Final Crystals. Composition Z is located within the composition triangle A-B-BC. When solidification is complete, the sample will be composed of crystals of A (10%), B (9%), and BC (81%).

2. Cooling Path of Melt. The melt will cool to point E where crystals of A, B, and BC are in equilibrium with melt. Point Z is located within the primary phase field of C, and consequently, crystals of C are the first to precipitate on cooling. The melt will change in composition from Z to point K located on the boundary line NP. The melt point then changes in composition from point K to point R. Simultaneously, crystals of C are being resorbed as indicated by a line drawn tangent to the boundary line NP[§] at point K and extending back to the corresponding Alkemade line BC-C. The intersection occurs at a point to the left of BC, which indicates that a *negative* quantity of C is precipitating; that is, C is being resorbed. The *overall* composition of the solid portion of the sample is indicated by the lower end of the tie line as it moves from point C toward point BC. (See Table 6.3.) When the lower end of the tie line reaches the point BC, all of the crystals of C will have been resorbed and the only crystalline phase present at this temperature is BC. The upper end of the tie line is at point R. Because BC is the only crystalline phase present, further cooling results in the composition of the melt moving directly across the primary phase field of BC as shown by the line RG. At point G, crystals of A appear and the melt composition moves along the boundary line from point G toward point E. Note that the tie lines which could be drawn within the temperature interval from G to E will all terminate on the Alkemade line A-BC. At a temperature just above the eutectic temperature (called E^+ in Table 6.3), crystals of BC and A are in equilibrium with melt of composition E. On further cooling, the melt will solidify in eutectic proportions: A = 30%, BC = 33%, and B = 37%.

§Boundary line NP is sometimes referred to as a "reaction boundary" because the melt changes in composition by reacting with crystals of C and forming crystals of BC.

Table 6.3. Calculation Sheet for Ternary Systems (See Fig. 6.29)

			Analysis		
Melt $\begin{cases} A=10 \\ B=25 \\ C=65 \end{cases}$		Final Crystals $\begin{cases} A=10 \\ BC=81 \\ B=9 \end{cases}$			

Temp.	Proportions and Composition of Phases		A	B	C
Z 47 ε	Melt=100% Crystals=ε% (C)	$\begin{cases} A=10 \\ B=25 \\ C=65 \end{cases}$	10	25	65
K 47	Melt=72%	$\begin{cases} A=14 \\ B=35 \\ C=51 \end{cases}$	10	25	37
$\frac{18}{65}$	Crystals=28% (C, BC)	$\begin{cases} C=100 \\ BC=\varepsilon \end{cases}$			$\frac{28}{65}$
R 20	Melt=42.5%	$\begin{cases} A=23 \\ B=32 \\ C=45 \end{cases}$	10	13.5	19
$\frac{27}{47}$	Crystals=57.5% (C, BC)	$\begin{cases} C=\varepsilon \\ BC=100 \end{cases}$ $\begin{cases} B=20 \\ C=80 \end{cases}$		11.5 $\frac{25.0}{65}$	$\frac{46}{65}$

Resorption of C xtals occurs from K to R

			A	B	C
G 20	Melt=30%	$\begin{cases} A=33 \\ B=37 \\ C=30 \end{cases}$	10	11	9
$\frac{46}{66}$	Crystals=70% (BC, A)	$\begin{cases} BC=100 \\ A=\varepsilon \end{cases}$ $\begin{cases} B=20 \\ C=80 \end{cases}$		$\frac{14}{25}$	$\frac{56}{65}$
E^+ 15	Melt=23%	$\begin{cases} A=30 \\ B=44 \\ C=26 \end{cases}$	7	10	6
$\frac{50}{65}$	Crystals=77% (BC, A)	$\begin{cases} \frac{132}{137}BC=96 \quad (74) \\ \frac{5}{137}A=4 \end{cases}$ $\begin{cases} B=20 \\ C=80 \end{cases}$	$\frac{3}{10}$	$\frac{15}{25}$	$\frac{59}{65}$

Melt solidifies to form eutectic microstructure: A,B,C(ℓ)$\rightleftarrows$A(s) + BC(s) + B(s)

			A	B	C
E^- 15	Eutectic=23%	$\begin{cases} A=30 \\ BC=33 \quad (7.5) \\ B=37 \end{cases}$ $\begin{cases} B=20 \\ C=80 \end{cases}$	7	1.5 8.5	6
$\frac{50}{65}$	Crystals=77% (BC, A)	$\begin{cases} \frac{132}{137}BC=96 \quad (74) \\ \frac{5}{137}A=4 \end{cases}$ $\begin{cases} B=20 \\ C=80 \end{cases}$	$\frac{3}{10}$	$\frac{15}{25}$	$\frac{59}{65}$

3. Cooling Path of Solid. As the melt composition changes from Z to K, crystals of C precipitate. During cooling to a temperature indicated by point R, the solid changes composition from point C to point BC, the C crystals are resorbed, and the quantity of BC is increased. At a temperature corresponding to point G, crystals of A begin to precipitate and the solid composition changes from point BC to point F as the sample cools to the eutectic temperature. During the eutectic reaction the solid composition changes from point F to point Z. These reactions are summarized graphically in Fig. 6.30.

6.9. Composition on an Alkemade Line

In Fig. 6.31 the composition K appears on the Alkemade line A-BC which does not intersect its boundary line PE, and therefore, the line A-BC

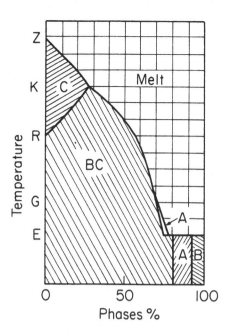

Fig. 6.30. Phase analysis diagram for composition Z of Fig. 6.29.

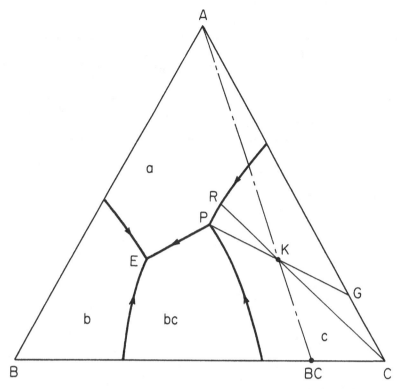

Fig. 6.31. Cooling path for melt of composition K located on Alkemade line A-BC.

does not represent a true binary system. Nevertheless, the melt will solidify to form A and BC in the proportions determined by the position of the point K on the Alkemade line, although the sequence of reactions is considerably different from that described previously for the case where the Alkemade line intersects its boundary (Section **6.4**). The analysis is as follows:

Isoplethal Study (Example 6.5)

1. Final Crystals. By application of the lever rule, the proportions of A and BC are found to be A = 30%, and BC = 70%.

2. Cooling Path of Melt. The melt will cool to the point at which crystals of A and BC are in equilibrium. This point is located at P, which is a peritectic point. Because the point K is located within the primary phase field of C, the first crystals to precipitate on cooling are those of C. The melt changes in composition from point K toward point R on the boundary line between the primary phase fields of A and C. As the melt changes composition along the boundary line from R to point P, crystals of A and C coprecipitate. At point P the peritectic reaction occurs in which the C crystals react with part of the melt to form crystals of A and BC:

$$C(s) + A,B,C(l) \leftrightarrows A(s) + BC(s)$$

The remainder of the melt crystallizes to form A and BC; thus when solidification is complete, the proportions of A and BC are those calculated above (A = 30%, BC = 70%). (See Table 6.4.)

3. Cooling Path of Solid. As the melt cools to point R, crystals of C precipitate. During cooling from point R to point P, the coprecipitation of A and C causes the composition of the solid to change from point C along the **Alkemade line AC to the point G. During the peritectic reaction and crystallization of the remaining melt, the solid changes in composition from** G to point K on the Alkemade line A-BC. These changes are summarized in the phase analysis diagram of Fig. 6.32.

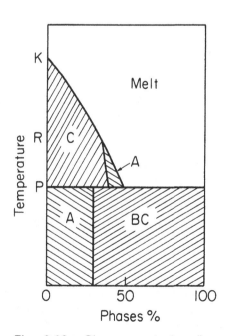

Fig. 6.32. Phase analysis diagram for composition K of Fig. 6.31.

97

Table 6.4. Calculation Sheet for Ternary Systems

Melt $\begin{cases} A=30 \\ B=14 \\ C=56 \end{cases}$ Final Crystals $\begin{cases} A=30 \\ BC=70 \end{cases}$

Temp.		Proportions and Composition of Phases		Analysis A	B	C
K	59	Melt=100%	$\begin{cases} A=30 \\ B=14 \\ C=56 \end{cases}$	30	14	56
	$\frac{\varepsilon}{59}$	Crystals=ε% (C)				
R	59	Melt=65%	$\begin{cases} A=46 \\ B=22 \\ C=32 \end{cases}$	30	14	21
	$\frac{32}{91}$	Crystals=35% (C, A)	$\begin{cases} A=\varepsilon \\ C=100 \end{cases}$			$\frac{35}{56}$
P^{+}	31	Melt=50%	$\begin{cases} A=40 \\ B=28 \\ C=32 \end{cases}$	20	14	16
	$\frac{31}{62}$	Crystals=50% (C, A)	$\begin{cases} A=20 \\ C=80 \end{cases}$	$\frac{10}{30}$		$\frac{40}{56}$
P^{-}	0	Melt=0%				
	All	Crystals=100%	$\begin{cases} A=30 \\ BC=70 \end{cases} \begin{cases} B=20 \\ C=80 \end{cases}$	30	14	56

4. *Consideration of Sections.* The sequence of reactions may be better understood by considering isothermal and vertical sections through the three-dimensional figure which represents the ternary system. Figure 6.33 shows a projection of the liquidus surface on which the isotherms have been drawn. The cooling path of the melt from 1100 to 800°C is shown on the diagram by the line KM. An isothermal section at 800°C is shown in Fig. 6.34. Note that the point K appears in the space of C crystals plus liquid and is on the tie line C-K-M. At 730°C, which is the peritectic temperature and the temperature of final solidification, the isothermal section appears as shown in Fig. 6.35. The point K now appears on the vertical plane A-BC (the projection of this plane on the liquidus surface is the Alkemade line) which separates the two three-phase spaces, A + BC + L and A + BC + C.

A vertical section through the three-dimensional figure along the Alkemade line A-BC is shown in Fig. 6.36. Note that the isopleth passes sequentially through the spaces C + L, A + C + L, and finally BC + A. It should be emphasized that this vertical section is not a true binary system, and therefore, valid tie lines cannot be constructed.

6.10. Phase Transformations

One or more of the crystalline phases in a ternary system may have other forms, or polymorphs, which are stable only within certain temperature ranges. The method of representing these different forms on a ternary diagram is shown in Fig. 6.37, where SiO_2 is one of the components of the system. Since a phase transformation occurs at a specific temperature under equilibrium conditions, the boundary line between two adjacent areas representing different polymorphs is coincident with the appropriate isotherm. For example, the boundary line between cristobalite and tridymite is coincident with the 1470°C isotherm.

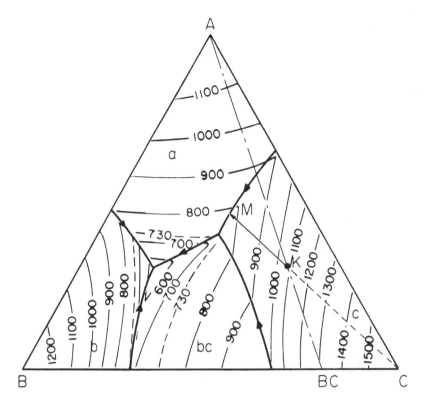

Fig. 6.33. Ternary system of Fig. 6.31, with isotherms added.

6.11. Decomposition of a Binary Compound Having a Phase Field in the Ternary System

In Fig. 6.38, the binary compound AB decomposes to A and B, which form a eutectic at a higher temperature. A primary phase field appears, however, in the ternary diagram because the liquidus surfaces slope

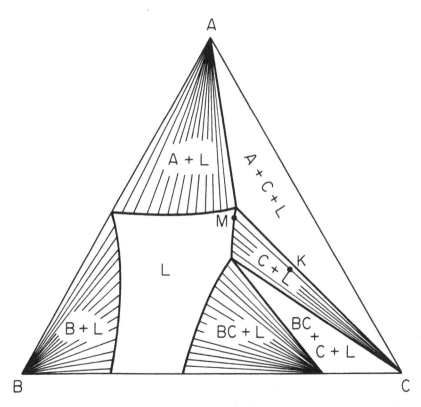

Fig. 6.34. Isothermal section at 800°C of ternary of Fig. 6.33.

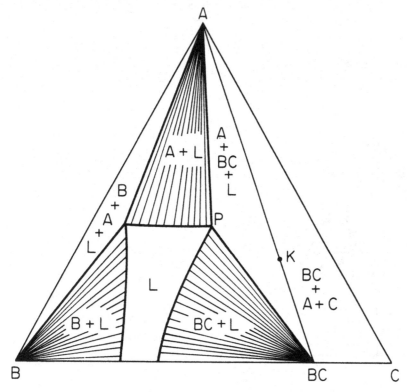

Fig. 6.35. Isothermal section at 730°C of ternary of Fig. 6.33.

downward to a temperature sufficiently low to intersect the space of AB, as shown in the lower part of Fig. 6.38.

This kind of system presents an interesting situation because the crystalline phases which are in equilibrium at point R (A, B, AB) do not fall within a composition triangle; therefore, no crystallization path will terminate at point R. The cooling path for a melt which passes through point R is shown in the diagram of Fig. 6.38. Resorption of crystalline A occurs during cooling from R to m.

6.12. Intermediate Ternary Compounds

When a ternary compound such as ABC in Fig. 6.39 melts congruently, the composition point representing the compound will occur within its primary phase field. In Fig. 6.39 the compound ABC forms three ternary subsystems with the components of the system. These subsystems, A-ABC-B, A-ABC-C, and B-ABC-C, are each independent ternary systems and could be studied as such.

If the compound ABC melts incongruently, its primary phase field will be separated from the composition point representing the compound, as shown in Fig. 6.40. In this case, the ternary subsystems are not true ternary systems.

6.13. Complex Cooling Paths

Most ternary ceramic systems involve several intermediate compounds, and consequently, the phase diagrams are usually more complex than those discussed thus far. Nevertheless, the sequence of reactions can be worked out by applying the principles used in the previous illustrative problems. For example, in Fig. 6.41, the composition X occurs in the composition triangle A-ABC-AB. The melt will cool to the point at which these three phases are in equilibrium (point e on the diagram). The cooling path is as follows: from point X to point m, C crystallizes from the melt. At point m, B crystals appear and coprecipitate with C as the melt cools to point s. At point s it would seem that there are two choices for the direction of the cooling path: one toward

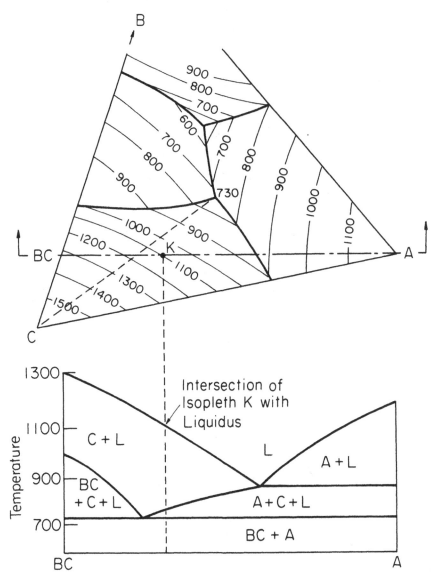

Fig. 6.36. Vertical section along Alkemade line *BC-A*.

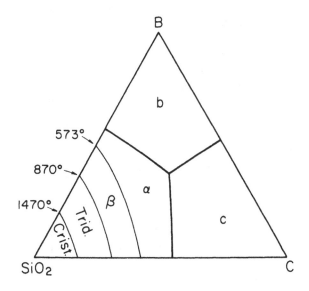

Fig. 6.37. Ternary diagram showing phase transformations of the SiO₂ component.

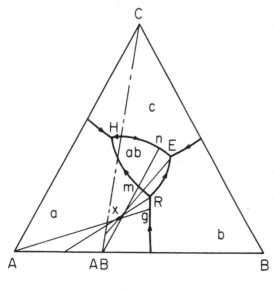

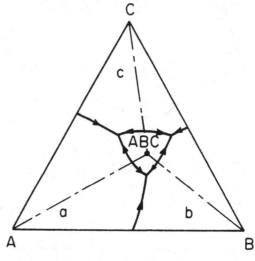

Fig. 6.39. Ternary compound ABC melts congruently.

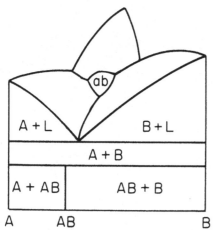

Fig. 6.38. Ternary system illustrating decomposition of a binary compound having a phase field in the ternary system. A cooling path for a melt of composition X is given by the sequence x-g-R-m-n-E.

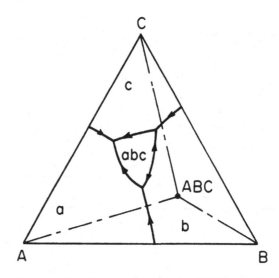

Fig. 6.40. Ternary compound ABC melts incongruently.

point f and the other toward point t. Examination of the path toward f shows it to be impossible, because as the melt composition changes from point s toward point f, the melt would have to be in equilibrium with crystals of ABC and C. These three phases (melt, ABC, and C) are at the apices of a triangle which does not include the original composition point X; thus, it would not be possible to form a sample of overall composition X from any conceivable mixture of crystals of ABC, crystals of C, and melt of compositions between s and f. Such is not the case with the path toward t, and so it is the path followed by the melt on cooling.

At point s a peritectic reaction occurs in which C crystals react with part of the melt to form crystals of ABC and B. When this reaction is completed, the tie line from point s through X terminates at its intersection with the Alkemade line ABC-B at point 3. Next, the melt cools along the boundary line st to point t, where another peritectic reaction occurs. B crystals react with a portion of the melt to form crystals of AB and ABC. At completion of the reaction, the tie line is from point t through X to point 5 on the Alkemade line AB-ABC.

The melt's composition subsequently changes along the boundary from t to point e, the eutectic point, and also the termination point for the cooling path of the melt. The last tie line is from point e through X to the point 6 on

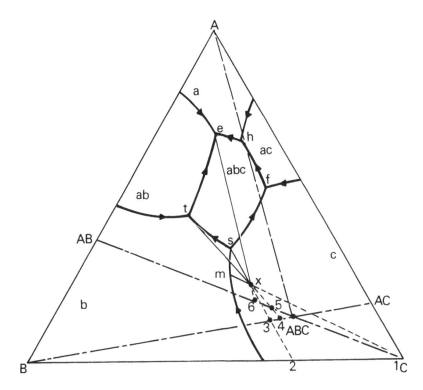

Fig. 6.41. Crystallization path for composition X in compatibility triangle AB-ABC-A.

the Alkemade line *AB-ABC*. During the eutectic reaction, the last portion of the melt solidifies in eutectic proportions.

During cooling, the solid portion of the sample follows the path *C-2-3-4-5-6-X*.

6.14. Ternary Solid Solution

In the ternary system shown in Fig. 6.42, there is complete solid solubility between components A and B to form the solid-solution phase α. In the binary system A-C, there is limited solubility of C in A but no solution of A in C. Similarly, in the binary system B-C, C is slightly soluble in B but B is not soluble in C. Thus, there is only one solid solution (phase α) in the ternary system, and its composition is seen to contain atoms of A, B, and C (except, of course, those compositions which fall on the line *AB*).

The binary systems B-C and A-C have eutectics at E_1 and E_2, respectively. These are joined by the three-phase boundary line E_1E_2. Along the line E_1E_2, liquid is in equilibrium with crystals of C and α. There is no ternary eutectic in the system. A plane projection of the liquidus surface is shown in Fig. 6.43.

The system is composed of six spaces: α, liquid, α + liquid, C + liquid, α + C + liquid, and α + C. An isothermal plane passed through the figure at 550°C intersects all of the six spaces, as shown in Fig. 6.44 and Fig. 6.45. In the two-phase regions, C + L and C + α, tie lines can be drawn from C to various points on the curves representing the compositions of the liquid and solid solutions, respectively. However, in the two-phase region, α + L, where both phases have variable composition, the tie lines must be determined experimentally.

The three-phase region, α + C + L, is called a *tie triangle*; it is composed of three tie lines. All composition points which fall within the triangle are composed of various proportions of the three phases represented by the apices of the triangle. The proportion of each phase is determined by the position of the composition point within the triangle.

103

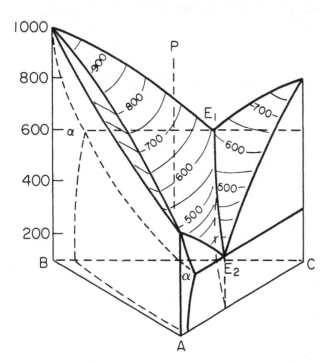

Fig. 6.42. Ternary system with solid solution.

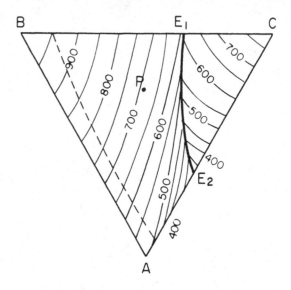

Fig. 6.43. Plane projection of liquidus surface of system shown in Fig. 6.42.

Within the two single-phase regions, α and L, compositions are determined in terms of A, B, and C.

In order to determine the cooling path for the melt or solid portion of a given sample, it is necessary to have isothermal sections in which the conjugate phases are shown by tie lines. The following isoplethal study will illustrate the use of such sections.

Isoplethal Study (Example 6.6)

The composition represented by the point P shown in Fig. 6.43 has been selected for analysis. Note that on cooling, this isopleth will intersect the liquidus surface at 700°C. At all temperatures above 700°C, the sample will consist of liquid only. For example, in Fig. 6.46 an isothermal section at 750°C shows the point P to be located in the area of liquid only.

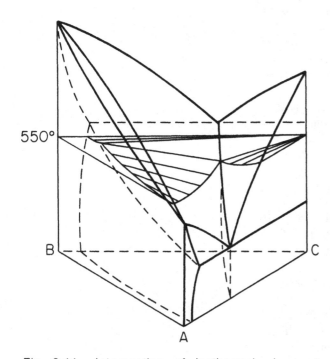
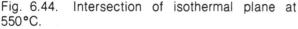

Fig. 6.44. Intersection of isothermal plane at 550°C.

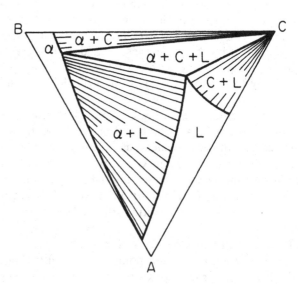

Fig. 6.45. Isothermal section at 550°C. Tie lines are drawn in two-phase region.

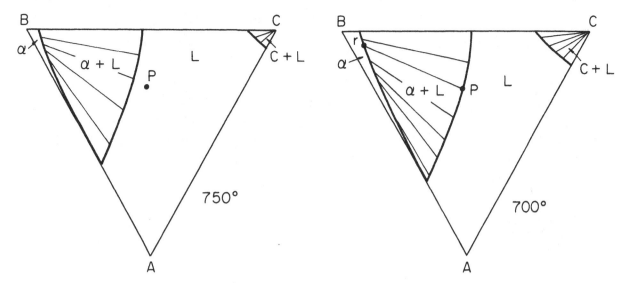

Fig. 6.46. Isothermal section at 750°C.   Fig. 6.47. Isothermal section at 700°C.

At 700°C (Fig, 6.47) the point P intersects the boundary of the $\alpha + L$ area. Thus, the sample is composed of liquid of composition P and an infinitesimal quantity of crystals of α solid solution of a composition given by the end of the tie line designated r in the diagram.

At 650°C (Fig. 6.48) point P intersects the space of $\alpha + L$ and falls on the tie line $2' - 2$ which gives the conjugate phases. The proportions of α and liquid are determined by applying the lever rule. For example, the percentage of liquid is given by $[(P - 2')/(2' - 2)](100)$, and the percentage of α is given by $[(P - 2)/(2' - 2)](100)$.

At 600°C (Fig. 6.49) the point P falls on the tie line $3' - 3$. Note that the compositions as well as the relative proportions of α and liquid are changing as the sample cools.

At 500°C (Fig. 6.50) the point P falls on the tie line $4 - 4'$. The composition of the melt lies on the three-phase boundary line E_1E_2; thus, an infinitesimal quantity of crystals of C will be present at this temperature in addition to α and liquid.

At 475°C (Fig. 6.51) the point P falls within the tie triangle $5' - C - 5$; therefore, $\alpha + C + L$ are the conjugate phases. Their proportions are determined by the position of point P within the triangle. For example, the tie line $5 - P - 5''$ can be constructed as shown. The percentage of melt of composition 5 is $[(5'' - P)/(5'' - 5)](100)$ and the proportion of the sample which is

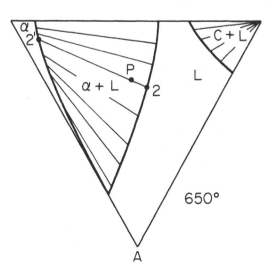

Fig. 6.48. Isothermal section at 650°C.

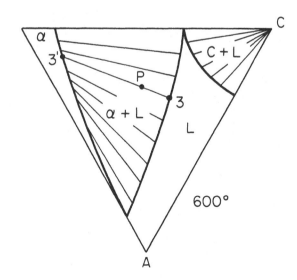

Fig. 6.49. Isothermal section at 600°C.

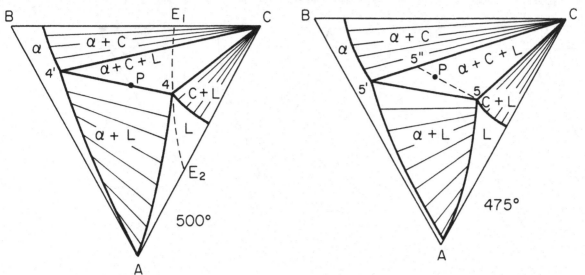

Fig. 6.50. Isothermal section at 500°C.

Fig. 6.51. Isothermal section at 475°C.

solid is $[(5 - P)/(5'' - 5)](100)$. The solid portion, however, is composed of α and C in the following proportions: % $\alpha = [(5'' - C)/(5' - C)](100)$, and the % C $= [(5' - 5'')/(5' - C)](100)$.

At 450°C (Fig. 6.52) the point P falls on the line $6' - C$, which is a side of the tie triangle $6'-C-6$. The sample is composed of α and C. The composition of the last portion of liquid to solidify is given by point 6. The proportion of α in the sample is $[(P - C)/(6' - C)](100)$, and the percentage of C is $[(6' - P)/(6' - C)](100)$. The sample is now completely solid.

These changes in composition of the solid and liquid portions of the sample during cooling are summarized in Fig. 6.53. The melt followed the path P-2-3-4-5-6 and the solid followed the path $1'$-$2'$-$3'$-$4'$-$5''$-P.

A vertical section through the system of Fig. 6.42 is shown in Fig. 6.54. The section was taken on a line from C through the point P to its intersection along the side BC. The position of the isopleth is shown by the dashed line.

6.15. Ternary System with Two Solid-Solution Phases

The system shown in Fig. 6.55 differs from the previous system in that a second solid-solution phase, β, is present. There are a total of seven spaces, one more than in the previous system. These spaces are shown in the 1600°C

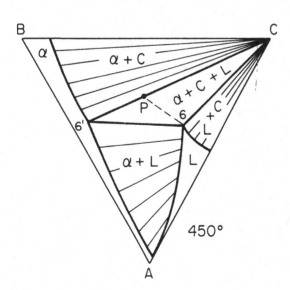

Fig. 6.52. Isothermal section at 450°C.

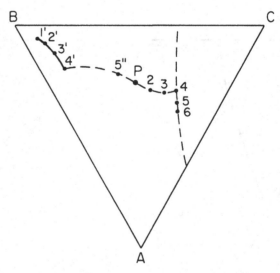

Fig. 6.53. Cooling paths of liquid and solid portions of sample of composition P.

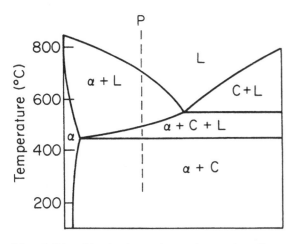

Fig. 6.54. Vertical section taken on a line from C through P of system shown in Fig. 6.42.

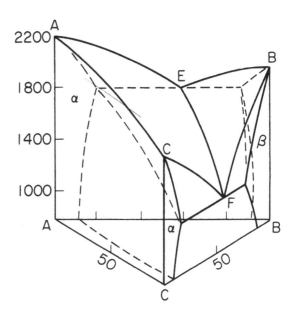

Fig. 6.55. Ternary system with two solid solution phases.

isothermal section of Fig. 6.56. Note that in all of the two-phase regions both phases have variable composition and therefore the tie lines must be determined experimentally.

6.16. System MgO·"FeO"·SiO$_2$

The system MgO·"FeO"-SiO$_2$ shown in Fig. 6.57 is a good example of a ceramic system involving extensive solid solution. This system has been studied in detail owing to its importance in steel plant refractories. An examination of the diagram shows several important refractories:

Periclase (MgO), mp 2825°C

Forsterite (2MgO·SiO$_2$), mp 1900°C

Silica (SiO$_2$), mp 1723°C

These refractory materials may be in contact with metallic iron, iron oxide, or silicate slags during the course of service. Thus, the diagram can be valuable in predicting or explaining refractory performance in steelmaking furnaces.

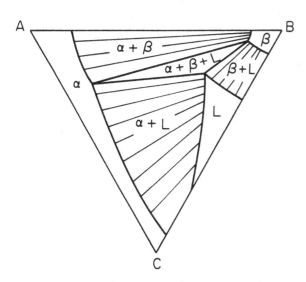

Fig. 6.56. Isothermal section at 1600°C.

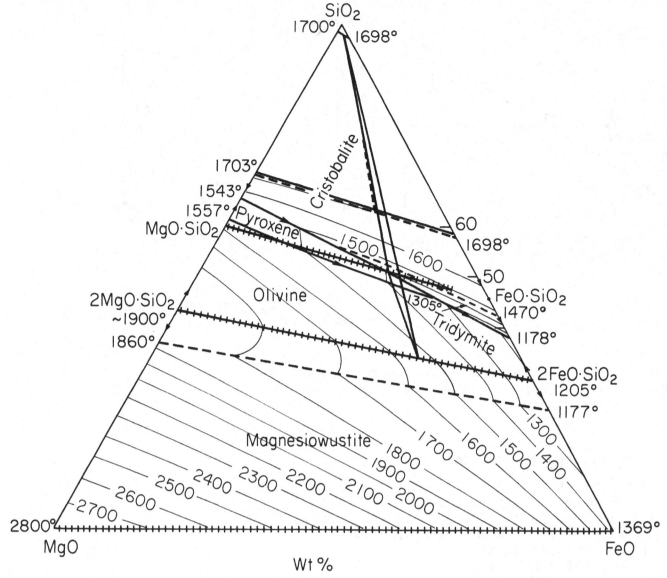

Fig. 6.57. System MgO-"FeO"-SiO₂ (oxide phases in equilibrium with metallic iron) (after E. F. Osborn and A. Muan, Phase Equilibrium Diagrams of Oxide Systems. American Ceramic Society, Columbus, Ohio, 1960).

Note that "FeO" (wustite) is set off in quotation marks. This is because FeO is not a stoichiometric compound. The phase wustite is a solid-solution phase which may exist over a range of iron contents less than the stoichiometric amount of iron represented by FeO. The Fe-O system of Fig. 6.58 indicates that at the exact FeO composition metallic iron and wustite coexist and that FeO is not experimentally obtainable (except at very low oxygen pressures).[1] The diagram (MgO-"FeO"-SiO₂) was determined, of necessity, under conditions which included the presence of metallic iron, and therefore, the diagram carries the modifying phrase, "oxide phases in equilibrium with metallic iron," beneath the title. For studies in which the total composition of the sample remains constant, the diagram accurately describes the equilibrium phases, although it is not truly a ternary system.

It is instructive to examine the binary systems which make up the ternary. In Fig. 6.59 the binaries are shown in relation to the ternary system. Complete solid solution exists between MgO and FeO to form the phase magnesiowustite, a solid-solution phase which consists of variable quantities of MgO, SiO₂, and FeO. Directly above the magnesiowustite field is a primary phase field designated *olivine.* The olivines are a solid-solution series, the end-members of which are forsterite (Mg₂SiO₄) and fayalite

[1]FeO has the rock salt (NaCl) crystalline structure in which some cation positions are vacant. The crystal must be considered as cation deficient, and on the phase diagram this is indicated as a phase higher in oxygen than the 1:1 ratio of FeO.

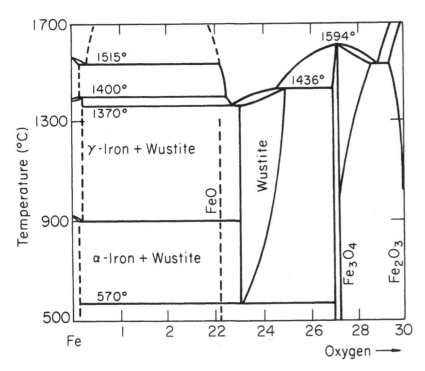

Fig. 6.58. System Fe-O_2 (after R. Hay and J. M. McLeod, *J. West Scotl. Iron Steel Inst.*, **52**, 109 (1944); P. T. Carter, M. Ibrahim, *J. Soc. Glass Technol.*, **36**, 144 (1952)).

(Fe_2SiO_4). The intermediate compositions may be written $2(Mg,Fe)O \cdot SiO_2$ or $(Mg,Fe)_2SiO_4$. The extent of solid solution between the two end-members is shown by the cross marks ($+++++++$) on the join between the end-members (Fig. 6.57). In this case, involving complete solid solubility, the cross marks extend the entire length of the join.

The *pyroxene* field is a small, triangularly shaped area adjacent to the olivine field. Structurally, the pyroxenes (or metasilicates) are composed of SiO_4 tetrahedra which share two oxygen atoms to form a single chain. The chains are cross-linked through divalent cations such as Mg, Fe, or Ca. Clinoenstatite ($MgO \cdot SiO_2$) forms one member of a solid-solution series which extends almost to the hypothetical compound $FeO \cdot SiO_2$. That is, Fe^{2+} can substitute for almost all of the Mg^{2+}.

Adjacent to the pyroxene field is the tridymite field. The boundary between tridymite and cristobalite is the 1470°C isotherm. The two-liquid region is delineated by shading along a line that extends from 1695°C on the $MgO\text{-}SiO_2$ join to about 1698°C on the $FeO\text{-}SiO_2$ join. At the very top of the diagram, the two-liquid boundary delineates a small cristobalite + liquid region which is part of the larger cristobalite + liquid region which extends under the immiscibility "dome".

There is one interior composition triangle or compatibility triangle which is made up of the following three Alkemade lines: one from the composition point SiO_2 to a specific pyroxene composition, another from this pyroxene composition to a specific olivine composition, and the last is from the olivine composition back to the SiO_2 composition point. Any melt composition falling within this triangle will cool to the point at which these three phases are in equilibrium. Examination of the diagram shows three fields intersect at a peritectic point at 1305°C. The isothermal section at 1305°C (Fig. 6.60) indicates the specific olivine and pyroxene compositions which are in equilibrium with the liquid.

Figure 6.61 shows a series of three isothermal sections at 1550, 1527, and 1450°C, respectively. The conjugate phases are indicated by the tie lines. It is

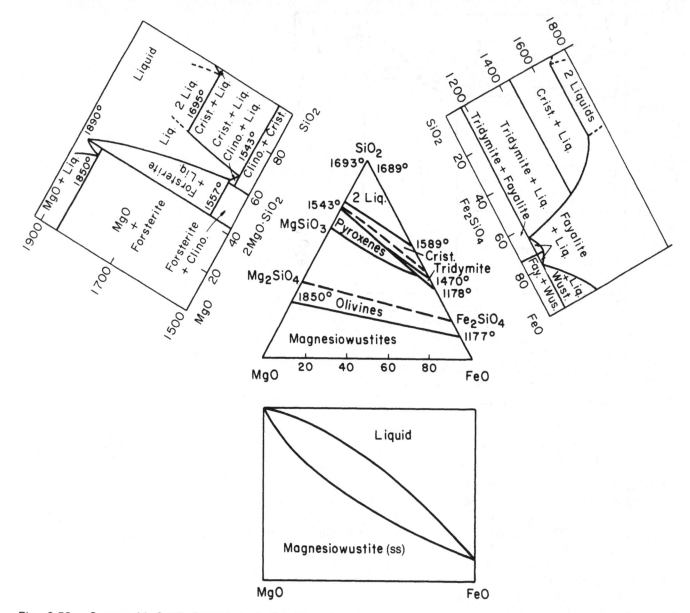

Fig. 6.59. System MgO-"FeO"-SiO₂ including binary systems.

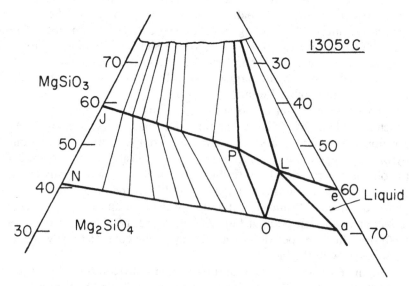

Fig. 6.60. Phase relations at 1305°C in system MgO-"FeO"-SiO₂.

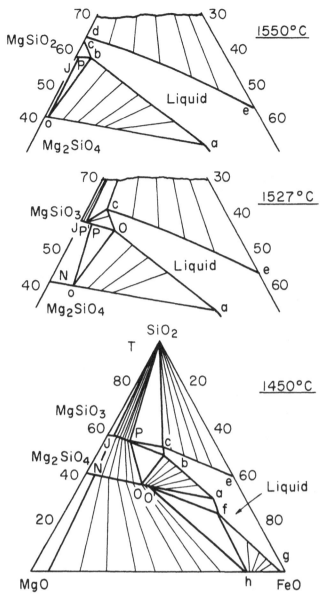

Fig. 6.61. Phase relations at 1550, 1527, and 1450°C
in the system MgO-"FeO"-SiO₂.

possible to estimate a crystallization path by utilizing the information on
these three diagrams.

Isoplethal Study (Example 6.7)

Consider a melt of composition X (Fig. 6.62) in the primary phase field
of olivine. The isopleth X intersects the liquidus surface at 1600°C, as indicated by its junction with the 1600°C isotherm. From the slopes of the tie
lines in the olivine + liquid areas of the 1550 and 1527°C isothermal sections,
it can be estimated that at 1600°C the slope of the appropriate tie line might
be as shown by the line $1 - X$ in Fig. 6.62. The slopes for suitable tie lines at
1550 and 1527°C can be taken from the appropriate sections and are shown
as the line 2-2', 3-3', and 4-4'. At 4', the cooling path for the melt intersects
the boundary line between olivine and pyroxene; on further cooling, pyroxene coprecipitates with olivine. The appropriate tie triangle (estimated from
the 1450°C isothermal section) is given by 4-4'-4''. With further cooling, the
tie line 6-6'' passes through the point X. The intersection of the tie line 6-6'
with the boundary line indicates that the temperature of final solidification is
approximately 1420°C. The solidified sample at that temperature will be
composed of a mixture of olivine and pyroxene of the composition given by

111

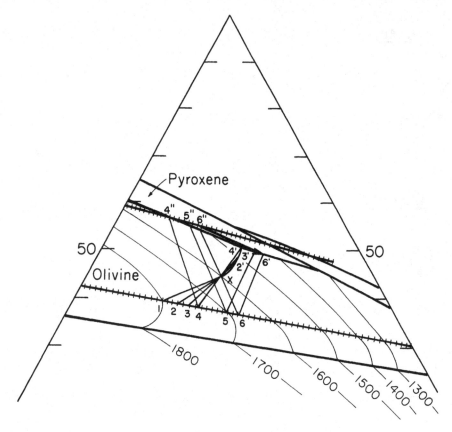

Fig. 6.62. Crystallization path of composition X estimated from slopes of tie lines in isothermal sections.

the extremities of the tie line 6-6″. The proportions of each may be estimated by application of the lever rule.

6.17. Liquid Immiscibility

Two-liquid regions in ternary systems occur in a number of ways. Frequently, the two-liquid region (or liquid miscibility gap) includes one or two

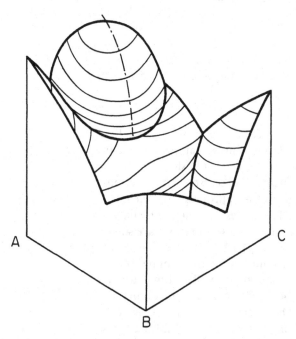

Fig. 6.63. Liquid miscibility gap in ternary system.

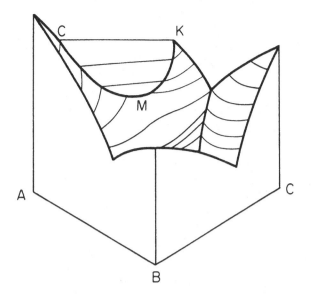

Fig. 6.64. Two-liquid region removed to show intersection with liquidus surface.

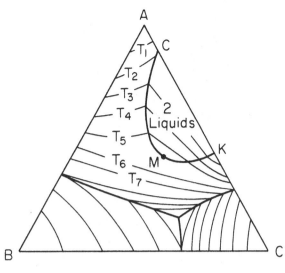

Fig. 6.65. Plane projection of liquidus surface (from Fig. 6.64).

of the binary sides; sometimes the two-liquid region is confined to the ternary field only. In Fig. 6.63, a two-liquid region extends from the binary side AC into the primary phase field of A. The intersection of this domelike space with the liquidus surface is shown in Fig. 6.64 where the dome has been removed to reveal the tie lines connecting the conjugate phases at several temperatures.

A plane projection of the liquidus surface of this system is shown in Fig. 6.65. An isothermal section at temperature T_4 (Fig. 6.66) reveals the triangular-shaped space of $A + L_1 + L_2$. In the space figure this space is generated by an isothermal line, one end of which slides down the A axis while the other end follows the line of intersection C-M-K. At point M, called the critical point, the composition of the two liquids becomes identical. An isoplethal study of a sample which undergoes liquid-liquid separation during cooling is as follows:

Isoplethal Study (Example 6.8)

Composition Y (Fig. 6.67) will cool to the eutectic point E, at which crystals of A, B, and C are in equilibrium with melt. At the temperature of intersection of the isopleth Y with the liquidus surface, crystals of A will precipitate and the melt composition will change along the line from Y toward t. At a temperature corresponding to that of point t, the liquid portion of the sample will begin to separate into two liquids. At a temperature corresponding to that at point 3, one liquid (L_1) will have a composition given by point d, and the other (L_2) that of point h. The proportions are calculated as follows:

At temperature 3

$$\% \text{ solid} = \frac{Y-3}{A-3}(100) \text{ (A crystals)}$$

$$L_1 = \frac{h-3}{h-d}(100) \left\{ \begin{array}{l} A \\ B \\ C \end{array} \right.$$

$$\% \text{ liquid} = \frac{A-Y}{A-3}(100)$$

$$L_2 = \frac{d-3}{h-d}(100) \left\{ \begin{array}{l} A \\ B \\ C \end{array} \right.$$

As the sample cools to temperature 4, L_1 changes in composition from point d to point e and L_2 changes from h to g.

At temperature 5, the composition of L_1 is given by point f and the composition of L_2 by point 5. The relative proportions of the two liquids have changed until at point 5 the quantity of L_1 is infinitesimal and the liquid portion of the sample is almost entirely L_2.

113

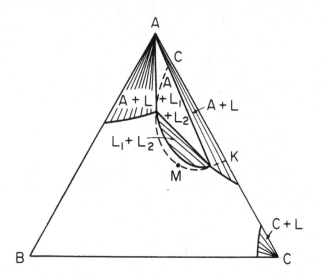

Fig. 6.66. Isothermal section at temperature T_4.

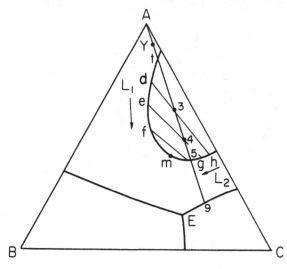

Fig. 6.67. Crystallization path for melt of composition Y.

When the sample cools from point 5 to point 9, only one liquid phase is in equilibrium with crystals of A. At point 9, C crystals coprecipitate with A as the melt cools along the boundary line toward point E, where final solidification takes place at the eutectic temperature.

When a liquid miscibility gap occurs in both binary systems which meet at a common point in the ternary system, the two-liquid region may extend completely across the diagram as shown in Fig. 6.68. The system MgO-"FeO"-SiO₂ , shown in Fig. 6.57, is an example of such an arrangement.

6.18. System MgO·Al₂O₃·SiO₂

Of considerable importance to the ceramic industry is the system MgO-Al₂O₃-SiO₂, shown in Fig. 6.69. A large number of commercial products can be represented on this diagram. The binary side Al₂O₃-SiO₂, which includes the refractories silica, alumina, and mullite, in addition to the fireclay refractories, has been discussed in Chapter 3. The binary side MgO-Al₂O₃ contains the compound MgO·Al₂O₃, which carries the mineral name spinel.** Magnesium aluminate spinel is used in various refractory applications and also finds use as a substrate material for integrated electronic circuits.

**The term "spinel" is also used to describe a specific crystal structure in which the oxygen atoms are in a face-centered cubic, close-packed arrangement. The cations are located in certain of the tetrahedral and octahedral interstices. Some common spinels are FeO·Cr₂O₃, MgO·Cr₂O₃, FeO·Fe₂O₃, and MgO·Fe₂O₃.

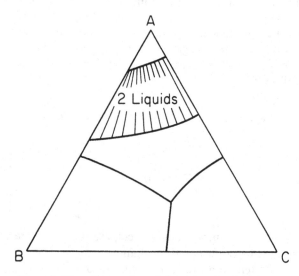

Fig. 6.68. Example of liquid miscibility gap extending continuously from binary A-B to binary A-C.

The binary side MgO-SiO₂ contains the refractory material periclase (MgO) and the compounds forsterite (2MgO·SiO₂) and enstatite (MgO·SiO₂). Forsterite is used as a furnace lining material owing to its volume stability and strength at high temperatures. Forsterite is also used as a low-loss electrical insulator. The low-temperature form of magnesium metasilicate is called enstatite and transforms to its high-temperature form, protoenstatite, at approximately 1040°C. On cooling, protoenstatite may transform to the metastable form called clinoenstatite. Magnesium metasilicate is the principal crystalline phase in the electrical insulators known as steatite ceramics, the compositions of which fall in the area labeled A in Fig. 6.69. The low-loss steatites have compositions in the range marked B in Fig. 6.69.

The ternary compound cordierite (2MgO·2Al₂O₃·5SiO₂) forms the primary crystalline phase of the cordierite ceramics whose compositions fall in the area marked C in Fig. 6.69. Cordierite ceramics, in addition to being good high-temperature electrical insulators, have excellent resistance to thermal shock as a consequence of their vary low thermal expansion coefficients.

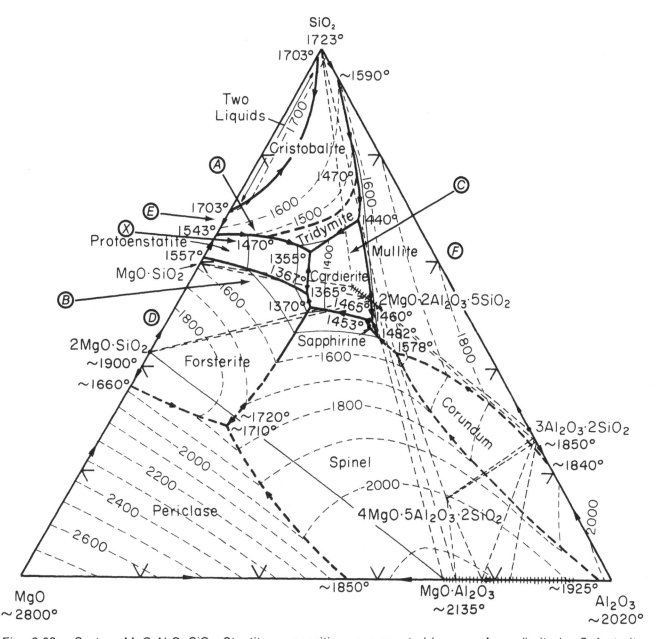

Fig. 6.69.   System MgO·Al₂O₃·SiO₂. Steatite compositions represented by area A, cordierite by C, fosterite ceramics by D. Point E corresponds to calcined talc and point F corresponds to calcined clay (after E. F. Osborn and A. Muan; No. 3 in Phase Equilibrium Diagrams of Oxide Systems. American Ceramic Society, Columbus, Ohio, 1960).

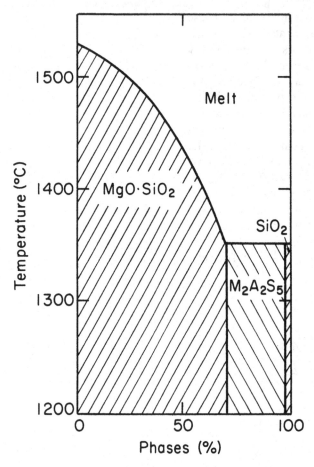

Fig. 6.70. Phase analysis diagram for steatite (composition X in Fig. 6.69).

These materials are often used as insulators in electric hot plates, as kiln furniture, and as burner nozzles.

The principal raw materials used in the manufacture of the cordierite and steatite ceramics are clay ($Al_2O_3 \cdot 2SiO_2 \cdot 2H_2O$) and talc ($3MgO \cdot 4SiO_2 \cdot H_2O$), the dehydrated forms of which are represented on the diagram of Fig. 6.69 as points F and E, respectively. The cordierite and steatite compositions are in areas of the phase diagram in which the liquidus surfaces are relatively shallow in slope; thus, the amount of liquid phase which is formed during the firing of these compositions changes rapidly with small increases in temperature. Consider the steatite composition labeled point X in Fig. 6.69; the phase analysis diagram for this composition is given in Fig. 6.70. Note that on heating to 1355°C, which is the temperature corresponding to that of the eutectic, approximately 25% melt would be expected to form. If the temperature were raised an additional 40°C, the amount of liquid phase would increase to about 40%. Since the optimum quantity of liquid for densification of this composition is 25–35%, careful control of the firing temperature would be required to prevent overfiring and subsequent distortion or bloating of the ware. The raw materials, talc, clay, and $MgCO_3$, commonly used in the manufacture of steatite ceramics usually contain small amounts of fluxes such as CaO, K_2O, and Na_2O, which favor the formation of liquid. The firing range of steatites and of cordierite ceramics is usually narrow owing to this relatively large increase in liquid with small increases in temperature.

6.19. System Na₂O·CaO·SiO₂

The high-silica region of the system Na_2O-CaO-SiO_2 (Fig. 6.71) is of particular interest to glass technologists. The compositions of the majority of the

116

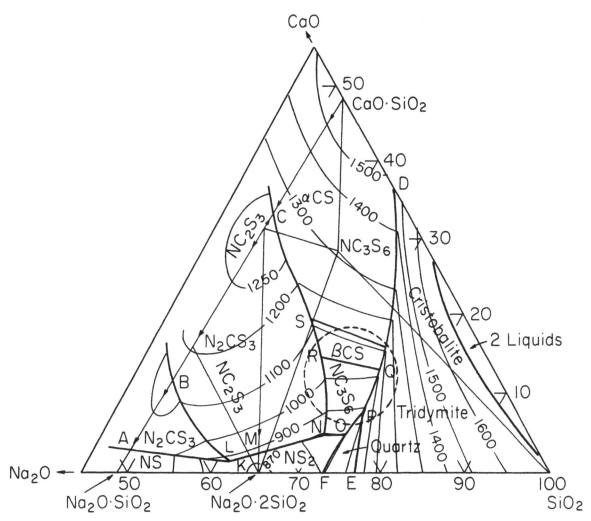

Fig. 6.71. High-SiO₂ corner of system Na₂O-CaO-SiO₂ (G. W. Morey and N. L. Bowen, *J. Soc. Glass Technol.*, **9** 232, 233 (1925)); dotted circle indicates approximate range of commercial soda-lime glass compositions.

commercial glasses, i.e., window, plate, and container glass, can be located in the area enclosed by the circle drawn on the phase diagram in Fig. 6.71. These compositions include the primary phase fields of devitrite $(Na_2O \cdot 3CaO \cdot 6SiO_2)$, beta wollastonite $(CaO \cdot SiO_2)$, and tridymite (SiO_2). Glass which is being cooled to a temperature suitable for forming may develop tiny crystals called glass stones. These crystals, undesirable from the point of view of the glass technologist, are generally the crystals which would be predicted from a consideration of the phase diagram. Changes in the cooling rate of the glass or small compositional adjustments may be required to avoid the formation of crystals.

Although the system Na_2O-CaO-SiO_2 is the basic phase diagram for commercial glass compositions, it should not be inferred that these glasses are three-component glasses. A number of commercial glass compositions are given in Table 6.5. The major constituents are Na_2O, CaO, and SiO_2, but there are frequently three and usually more minor constituents which are

Table 6.5. Representative Commercial Glass Compositions

Type of glass	SiO₂	CaO	MgO	Na₂O	K₂O	Al₂O₃
Container	72	3	7.5	15	1	1.5
Window	71	10	3	14	1	1
Plate	71	11	4	13		1

needed to give the glass the desired workability or to provide the required end-use properties. It is possible to arrange the major oxides into three groups, alkalies, alkaline earths, and glass formers, and thus reduce the composition to essentially a ternary system. For example, the Na_2O and K_2O can be grouped and treated as Na_2O; the CaO and MgO may be totaled as CaO. This treatment is justifiable because the phase diagrams of the systems $CaO-SiO_2$ and $MgO-SiO_2$ are very similar in the high-silica region. The systems Na_2O-SiO_2 and K_2O-SiO_2 are also quite similar in their respective high-silica regions.

Problems for Chapter 6

6.1. For Fig. 6.72.
   a. Label primary phase fields.
   b. Construct Alkemade lines.
   c. Indicate slopes of boundary lines.

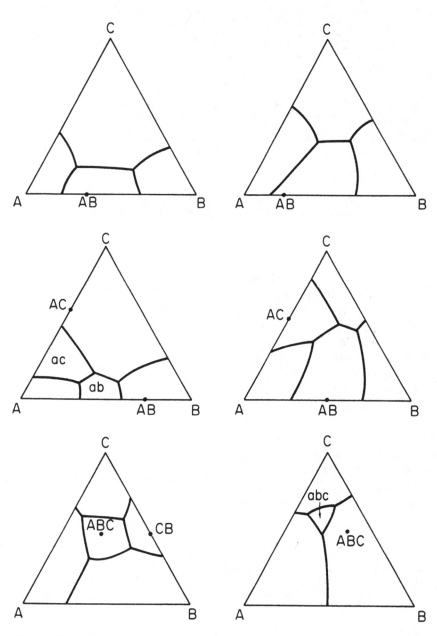

Fig. 6.72. Illustrations for problem 6.1.

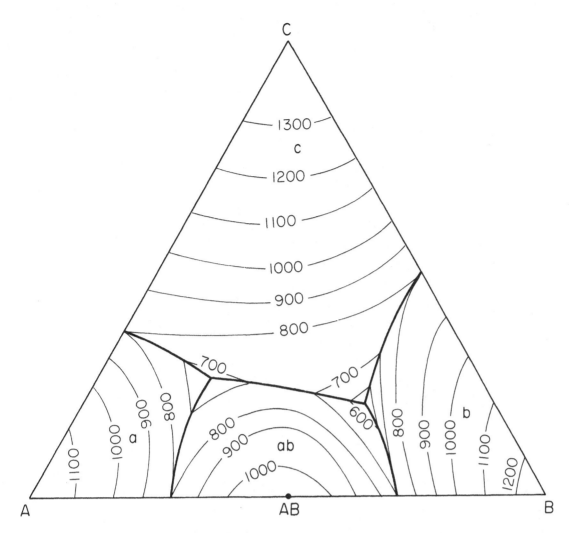

Fig. 6.73.   System for problem 6.2.

6.2. For Fig. 6.73.
   a. *Make an isoplethal study for a melt of composition* A = 30%,
      B = 5%, C = 65%.
   b. *Make a sketch of the isothermal section at 1100°C and at 700°C.*
   c. *Make a sketch to represent the vertical section taken along a line
      from C to 80% B (on the line* AB).

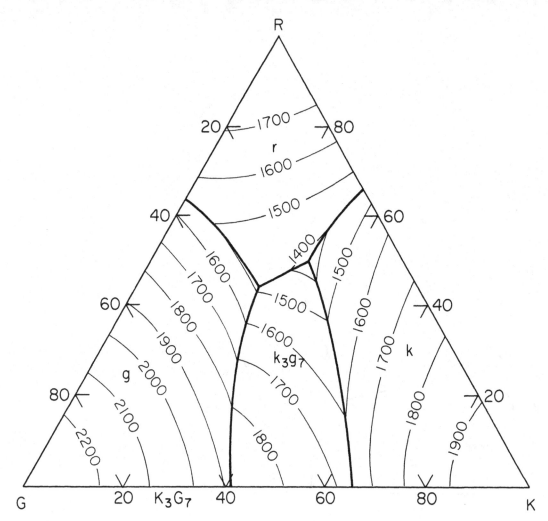

Fig. 6.74. System for problems 6.3 and 6.4.

6.3. *For Fig. 6.74.*
  a. *Make an isoplethal study for a melt of composition G = 70%, R = 10%, K = 20%.*
  b. *Construct a phase analysis diagram from the data calculated in part (a).*

6.4. *For Fig. 6.74.*
  a. *Make isoplethal studies for melts of the following compositions:*
    Melt 1: K = 35%, G = 60%, R = 5%.
    Melt 2: K = 10%, G = 20%, R = 70%.
  b. *Construct phase analysis diagrams for the compositions listed in part (a).*

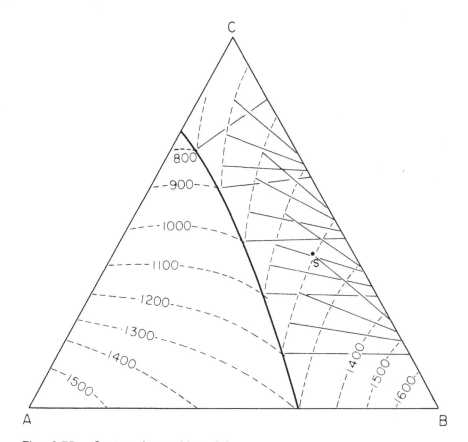

Fig. 6.75.   System for problem 6.5.

6.5. *In Fig. 6.75, complete solid solution occurs between components B and C. Systems A-B and A-C are binary eutectic systems. The dashed lines are isotherms. The solid lines are tie lines.*

  a . Determine the temperature of final solidification for a melt of composition "S".

  b . Indicate the approximate cooling path for the melt of composition "S".

  c . Determine the proportions and compositions of the phases present below the temperature of final solidification.

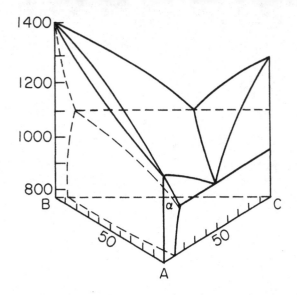

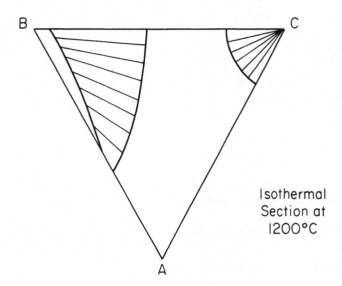

Fig. 6.76. System for problem 6.6.

6.6. *For Fig. 6.76.*
   a. Label all areas of the 1200°C isothermal section.
   b. Determine the proportion and composition of the phases for a melt of A = 15%, B = 70%, C = 15%, which has been cooled to 1200°C.

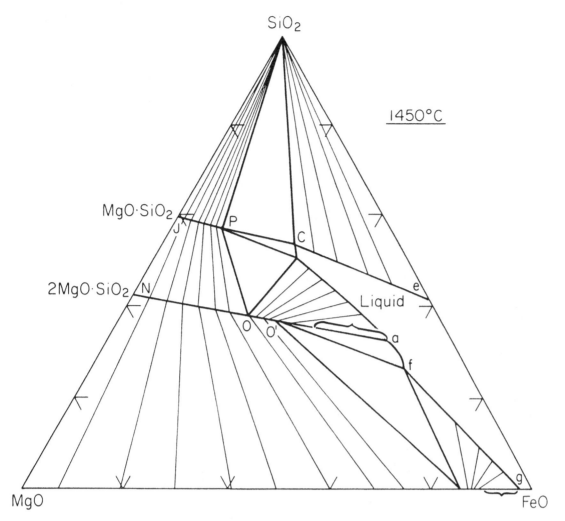

Fig. 6.77. System MgO-"FeO"-SiO₂, isothermal section at 1450°C, for problem 6.7.

6.7. *For Fig. 6.77 and 6.57.*
    A melt of composition SiO₂ = 40%, MgO = 30%, FeO = 30% is cooled
    to 1450°C under equilibrium conditions. Determine the following:
    a. The phases present.
    b. Their proportions.
    c. The composition of each phase.

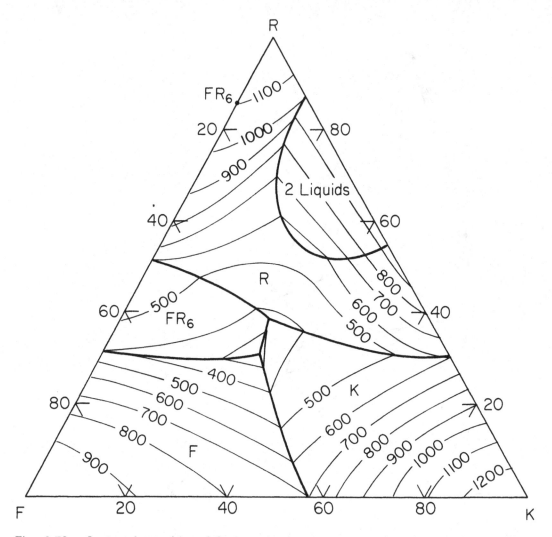

Fig. 6.78. System for problem 6.8.

6.8. *For Fig. 6.78.*

    a. Indicate the cooling path for a melt of composition R = 70%, F = 10%, K = 20%.

    b. Calculate the percent of each phase present at 700°C.

    c. Calculate the composition of each phase present at 700°C.

Bibliography and Supplementary Reading

N. L. Bowen and J. F. Schairer, *Am. J. Sci.*, **29**, 151 (1935).

P. T. Carter and M. Ibrahim, *J. Soc. Glass Technol.*, **36**, 144 (1952).

R. A. Giddings and R. S. Gordon, *J. Am. Ceram. Soc.*, **56** [3] 111 (1973).

R. Hay and J. M. McLeod, *J. West Scotl. Iron Steel Inst.*, **52**, 109 (1944).

J. D. Hodge and H. K. Bowen, *J. Am. Ceram. Sci.*, **65** [11] 582 (1982).

M. L. Keith and J. F. Schairer, *J. Geol.*, **60**, 181 (1952).

G. Masing; Ternary Systems, Reinhold, New York, 1944.

G. W. Morey and N. L. Bowen, *J. Soc. Glass Technol.*, **9**, 232 (1925).

A. Muan and E. F. Osborn; Phase Equilibria Among Oxides in Steel Making. Addison-Wesley, Reading, Mass., 1965.

E. F. Osborn and A. Muan; Phase Equilibrium Diagrams of Oxide Systems, Plate 3. The American Ceramic Society, Columbus, Ohio, 1960.

E. F. Osborn and M. Muan; Phase Equilibrium Diagrams of Oxide Systems, Plate 8. The American Ceramic Society, Columbus, Ohio, 1960.

G. A. Rankin and H. E. Merwin, *Am. J. Sci.*, **45**, 301 (1918).

K. A. Shahid and F. P. Glasser, *Phys. Chem. Glasses.*, **12**, 50 (1971).

D. R. F. West; Ternary Equilibrium Diagrams, 2d ed. Chapman and Hall, New York, 1982.

Stable phase equilibrium diagrams represent the phases that may be expected from reactions that occur under equilibrium conditions and, therefore, provide a basis for making predictions of the behavior of materials under various conditions of service or processing. Not all reactions reach complete equilibrium, and consequently, the phases which are present in a given system may not be the equilibrium phases. These departures from equilibrium may occur for a number of reasons. For example, there may have been insufficient time for the reactions to go to completion, or intervening metastable phases may have formed at some time during the process. Siliceous melts are, for instance, extremely viscous and reactions involving these melts occur very slowly. Recent developments in unusual ceramic processing techniques involving rapid solidification, sol-gel powder preparation, plasma spraying, and vapor condensation of films have made it all the more possible to attain nonequilibrium conditions in ceramics. In this chapter we examine some typical compositional and processing approaches and reactions that lead to nonequilibrium phases.

7.1. Reactions during Heating

Consider the reactions which occur during the heating of a mixture of A and B in the proportions shown by X_0 on the phase diagram of Fig. 7.1. In typical reactions, the raw materials A and B would be in powder form. After carefully mixing the raw materials, they might be reacted in a crucible or, possibly, pressed into some shape prior to heating. In either case the microstructure before heat treatment would appear approximately as shown schematically in Fig. 7.2. The particles of A and B are randomly mixed and the void space accounts for 30–50% of the total volume.

When the mixture is heated to a temperature just below T_1, no reactions would be expected except for possibly a small degree of solid-state sintering (densification). When the mixture is heated to temperature T_2, the sample would be expected, when equilibrium is established, to consist of crystals of A in a matrix of melt of the composition given by X_1. The first liquid to form on heating to a temperature above that of the eutectic (T_1) will occur at the contact points between particles of A and B, as shown in Fig. 7.3, and would approximate the composition of the eutectic ratio.

The portion of the melt that is in contact with crystals of component A (Fig. 7.3 (b)) will dissolve additional amounts of A, thus becoming enriched

(a)

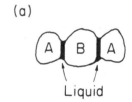

(b)

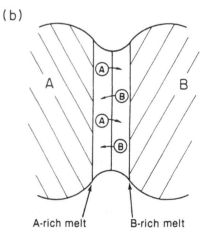

Fig. 7.3. (a) Liquid forms at contact points during heating. (b) Counter-diffusion of A and B atoms occurs within liquid phase due to composition gradient in the melt.

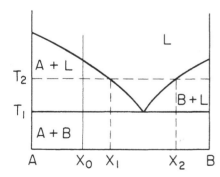

Fig. 7.1. Example of an A-B mixture (overall composition X_0) heated to temperature T_2.

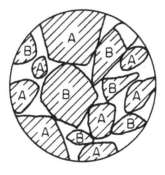

Fig. 7.2. Microstructure of an unfired mixture of A and B.

in A and causing the melt to approach the equilibrium concentration of A (as shown by the composition X_1 on the phase diagram of Fig. 7.1. Once the equilibrium concentration of A has been reached, further dissolution of A would cease.

Similarly, the melt in contact with particle B will become richer in B as more B is dissolved. Dissolution of B will cease when the composition of the melt reaches the equilibrium concentration given by X_2 in Fig. 7.1.

The simultaneous dissolution of A at one surface of the melt and that of B at the other surface produces a composition gradient in the melt. This gradient provides the driving force for the counterdiffusion of A and B atoms within the melt (Fig. 7.3(b)). The counterdiffusion of A and B atoms prevents the A-rich and B-rich melts from reaching their saturation concentrations; consequently, the dissolution process continues until all of the B is consumed, after which the system consists of crystals of A in a melt of composition given by point X_1 in Fig. 7.1. Depending upon the fluidity of the melt, the approach to equilibrium could be very slow. If the sample of composition X_0 were not heat-treated for a period of time sufficient to attain equilibrium, the resulting microstructure could contain some crystals of B as well as a smaller amount of liquid phase than that predicted by the phase diagram.

7.2. Suppression of an Intermediate Compound

The ease with which a crystalline phase may be nucleated is dependent upon a number of interrelated factors such as the free energy difference between the crystal and melt, the liquid-solid interfacial energy, the complexity of the crystal structure, and the degree of association of the melt. These factors may be different for each crystal-melt combination considered and can result in marked differences in nucleation behavior among crystalline phases within the same system. Consider a melt of composition X in the phase diagram of Fig. 7.4. On cooling to point K, crystals of AB would precipitate

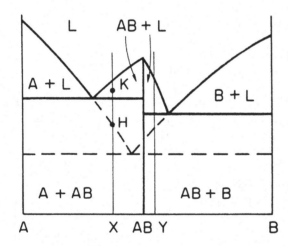

Fig. 7.4. Possible metastable version of the system is shown by dashed lines.

under equilibrium conditions. If however, the crystalline phase AB is not readily nucleated, the melt may supercool to point H (the intersection of isopleth X with the metastable extension of the liquidus line for crystalline A), at which temperature crystals of A may nucleate and grow. A similar situation may obtain for composition Y, in which B crystals may form rather than crystals of AB. The metastable equilibrium diagram might thus indicate a simple eutectic system as shown by the dashed lines of Fig. 7.4.

A relevant example of the metastable equilibrium conditions realized under suitable conditions in the system SiO_2-Al_2O_3 is shown in Fig. 7.5. The stable liquidi of SiO_2 and Al_2O_3 and the compound mullite form a stable phase diagram as shown in an earlier chapter (Fig. 4.2). When alumina-rich melts (~78–80 wt%) are slowly solidified below the equilibrium peritectic

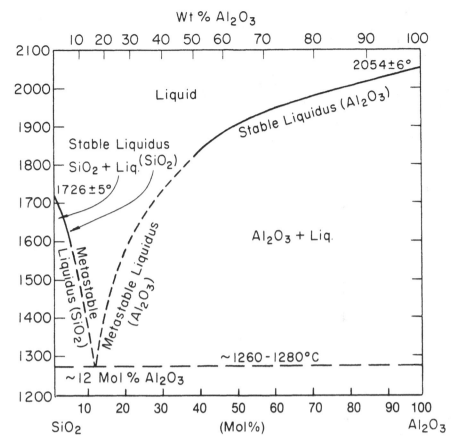

Fig. 7.5. Stable and metastable liquidi of SiO_2 and Al_2O_3 in the binary system Al_2O_3-SiO_2 (after I. A. Aksay and J. A. Pask, *J. Am. Ceram. Soc.,* **58** [11–12] 507–12 (1975); and S. H. Risbud and J. A. Pask, *J. Mater. Sci.,* **13,** 2449 (1978)) The stable liquidi and the stable phase diagram involve equilibria which also contain the intermediate compound mullite (see Fig. 4.2). The metastable liquidi and the simple eutectic (~12 mol% Al_2O_3; 1260°-1280°C) metastable phase equilibrium diagram shown is supported by experimental data on solidified high-Al_2O_3 melts (F. L. Kennard, R. C. Bradt, and V. S. Stubican; p. 580 in Reactivity of Solids. North-Holland, Amsterdam, 1972) and lower temperature (~1200–1300°C) liquid-phase formation in fired cristobalite-corundum powder compacts.

temperature ($\sim 1828°C$), the precipitation of Al_2O_3 crystals is observed in a liquid matrix (supercooled); the formation of mullite is suppressed. Continued cooling results in a metastable equilibrium between Al_2O_3 crystals and a liquid of variable composition (shown by the dashed lines in Fig. 7.5). The final metastable phases then consist of Al_2O_3 and the supercooled liquid that has frozen into a glass.

7.3. Glass Formation and Transformation Curves

We have discussed examples of metastable crystalline phases formed under conditions that represent some departure from equilibrium. In what must be considered the ultimate case of nonequilibrium, a liquid or a vapor phase can be formed into a metastable glassy or amorphous state. Differences in terminology exist in that the quenched phases are variously described as glasses, noncrystalline solids, or amorphous materials, depending on the background (and bias!) of the person studying these phases. Much of the technology related to *bulk* glasses and glass-ceramics, however, involves cooling a melt from above the liquidus temperature of the mixture to room temperature without the formation of a crystalline phase. Commercial glass compositions are selected in such a way as to take advantage of the tendency of many siliceous melts to supercool to a metastable liquid, which may subsequently be cooled to the solid material called glass.

127

In order for a crystal to form within a supercooled melt, a nucleus composed of atoms in crystalline array must be formed and must be of sufficient size to be thermodynamically stable. The probability of a sufficient number of atoms of suitable energy and orientation spontaneously coming together within the melt to form a stable nucleus is very low at small undercoolings. However, foreign surfaces within the melt may serve as nucleation sites and will reduce the number of atoms required for a stable nucleus. Surfaces such as container walls, bubbles, the interface between an immiscible liquid phase and its liquid matrix, or foreign material within the melt frequently serve as nucleation sites for crystal growth.

In many oxide melts and especially in siliceous melts, the rate at which stable nuclei are formed, even in the presence of foreign surfaces, is very low owing in large measure to the low mobility of the atoms and the associated atomic groups within these very complex melts.

Consider the example of a melt of composition Y_0 shown in Fig. 7.6.

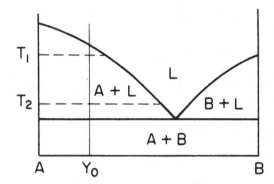

Fig. 7.6.   Melt Y_0 may be supercooled to temperature T_z if nucleation of A is avoided.

When this melt is cooled to temperature T_1, it becomes supersatuated in A and a small quantity of crystalline A precipitates from the melt under equilibrium conditions. If there are no stable nuclei present, crystallization of A will not occur and the melt is said to be in a metastable condition. With further cooling, such as to T_2 in Fig. 7.6, the degree of supersaturation becomes greater, the tendency (thermodynamic driving force) for crystallization becomes greater and the probability of forming stable nuclei is enhanced.

Acting in the opposite direction, the viscosity of the melt increases rapidly with decreasing temperature, and consequently, the mobility of the atomic species within the melt is decreased, thereby reducing the rate at which stable nuclei can be formed. These two opposing forces, i.e., the greater thermodynamic driving force for nucleation and the decreasing atomic mobility, give rise to a relatively narrow temperature range over which crystallization is likely to occur on cooling a melt. Fairly rapid cooling through this critical temperature range precludes crystallization and a glassy or vitreous phase is formed.

Much effort has been made in recent years to develop an understanding of factors that promote glass formation in ceramics and other materials. There is growing recognition that bulk glasses in large sizes or noncrystalline solids in relatively thin configurations can be formed from chemical mixtures covering the entire spectrum of bonding type. The specific processing for a particular material may vary from simply freezing a silicate liquid at normal furnace cooling rates to rapidly impacting a molten alloy on a rotating cold substrate to form thin amorphous tapes (splat cooling). Structural and thermodynamic principles have been historically invoked to explain glass formation, but modern treatments based on kinetic considerations advanced by

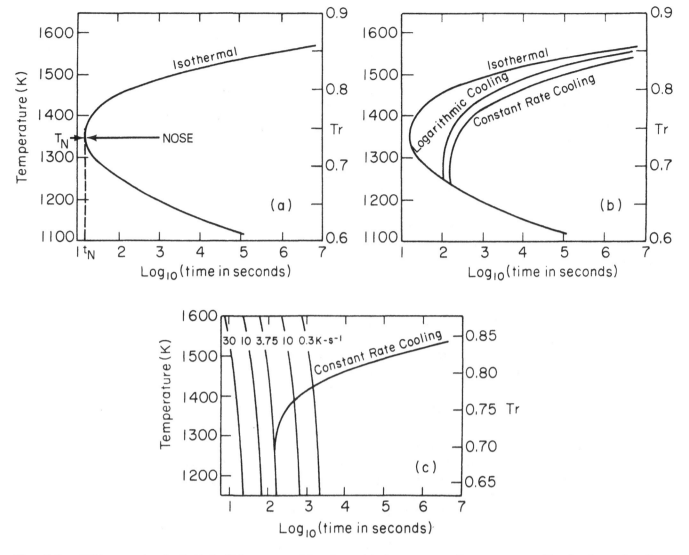

Fig. 7.7.   *TTT* curve for CaO-Al$_2$O$_3$-SiO$_2$ composition (anorthite) corresponding to a crystallized volume fraction of 10^{-6} (*a*). *CT* curve for anorthite (*b*) and cooling rate curves superimposed on *CT* curve (*c*), establishing 3.75 K/s as the critical cooling rate for glass formation in anorthite (after D. R. Uhlmann, *J. Am. Ceram. Soc.*, **66** [2] 95–100 (1983)).

Uhlmann and his colleagues (see Bibliography) provide a far more general view of glass formation in *any* material. Thus, considerations of viscosity and cooling rates imposed on a melt, presence of nucleating heterogeneities, nucleation and growth kinetics, and solute redistribution are all factored in to develop an understanding of glass formation in terms of time-temperature-transformation (*TTT*) and continuous cooling (*CT*) curves. Typical examples of a *TTT* curve for a CaO-Al$_2$O$_3$-SiO$_2$ composition are reproduced in Fig. 7.7(*a*), and the continuous cooling (*CT*) curves for the same composition are shown in Fig. 7.7(*b*). The *TTT* curve of Fig. 7.7(*a*) corresponds to an extremely small amount of crystalline material (volume fraction crystallized $\approx 10^{-6}$) and can be used to estimate a critical cooling rate for glass formation, $(dT/dt)_c$, given as

$$\left(\frac{dT}{dt}\right)_c = \frac{(T_E - T_N)}{t_N} \tag{7.1}$$

where $T_E - T_N$ and t_N = undercooling and time at the nose of the *TTT* curve, respectively (Fig. 7.7(*a*)). Time t_N is also used to estimate the thickness, y_c, that can be obtained as a glass: $y_c = (D_{th}t_N)^{1/2}$; D_{th} = thermal diffusivity of ma-

129

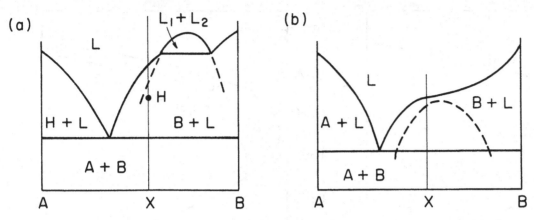

Fig. 7.8. Metastable liquid immiscibility in binary systems: (a) dashed lines indicate metastable extension of stable two-liquid region; (b) dashed lines indicate region of metastable liquid immiscibility.

terial in cm²/s. The CT curve and cooling rate curves as shown in Fig. 7.7(c) can be used to establish a critical cooling rate for glass formation. The critical cooling rate is given by the cooling curve that just misses intersecting the CT curve (3.75 K/s as shown in Fig. 7.7(c)).

7.4. Metastable Immiscibility

In many systems of interest to ceramists, there are compositional regions which exhibit liquid miscibility gaps such as shown in Fig. 7.8(a). Frequently these two-liquid regions extend metastably to temperatures below that of the liquidus. In some instances (Fig. 7.8(b)) the two-liquid region appears as a metastable region only.

If melts in these systems are supercooled into the metastable two-liquid region as in the case of composition X of Fig. 7.8(a and b), the liquid may separate into two liquids as described earlier in Chapter 3.

The liquid-phase separation may occur by classical nucleation and growth or by spinodal processes depending on the original melt composition, cooling rate, and other factors. The equilibrium crystalline phase may subsequently nucleate at the energetically favorable boundary between the two liquid phases. This mechanism of nucleating crystalline phases within a glass is important to the production of the class of materials called "glass-ceramics," in which an object is formed in the glassy state and subsequently converted to the crystalline state by means of a heat treatment in which the glass is nucleated and then brought to a higher temperature where crystal growth may occur. Although liquid-liquid separation is an important nucleation mechanism in "glass-ceramics," it should be noted that nucleation in glass-forming systems may occur by other mechanisms as well.

7.5. Solid Solutions

Cooling a sample through a two-phase region consisting of a solid solution and a liquid involves a continuous change in the compositions of both the solid and the liquid phases. Diffusion in the solid state is very slow, and thus, extremely slow cooling may be required to maintain equilibrium.

Consider a melt of composition X in the binary system A-B (Fig. 7.9) in which the components are completely miscible in both the liquid and solid states. Under equilibrium cooling conditions, the first crystals to precipitate from the melt will have the composition α_1 and the liquid will be of composition L_1. When the sample is cooled to temperature T_2, the overall composition of the crystalline phase will change from α_1 to α_2 (B atoms will diffuse into the existing crystals and A atoms will diffuse outward in order to accomplish this compositional change); similarly, the liquid composition will

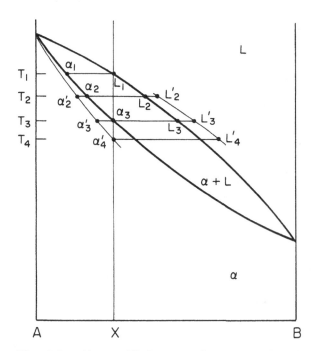

Fig. 7.9. Nonequilibrium cooling of melt of composition X.

change from L_1 to L_2. Final solidification will take place at temperature T_3; the last melt to solidifiy will have the composition L_3 and the crystals that of α_3.

If the melt is cooled too quickly for equilibrium conditions to be maintained, the crystalline phase will develop a composition gradient (the liquid phase will also develop a compositional gradient), but owing to the faster diffusion of atoms in the liquid than in the solid portion of the sample, the gradient in the liquid is considerably less than that in the solid. A schematic drawing of the microstructural and compositional changes which occur during cooling is shown in Fig. 7.10. As the sample cools from temperature T_1 to T_2, the additional crystalline material either precipitates on the existing crystals of α_1 or forms new crystals if conditions for nucleation are favorable.

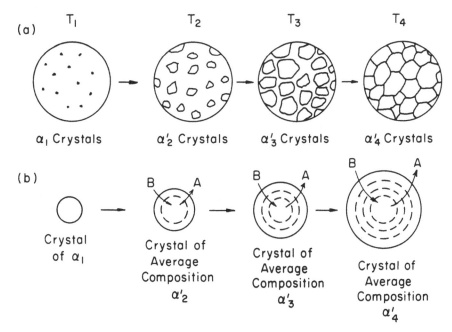

Fig. 7.10. Changes in (a) microstructure and (b) composition during nonequilibrium cooling of melt X in the system of Fig. 7.9.

It is more probable that most of the growth will occur on the existing crystals. The outside layers of the crystal become progressively richer in B as the system attempts to reach the equilibrium compositions of the solid and liquid corresponding to temperature T_2. Concurrently, there is a diffusion of B atoms into the interior and a diffusion of A atoms toward the exterior owing to the compositional gradient in the crystal. Because equilibrium is not attained, a compositional gradient remains and the average composition of the crystalline phase is deficient in B as shown by α_2'. The average composition of the liquid phase will be correspondingly slightly richer in B than predicted by the equilibrium phase diagram, as shown by L_2' (if the system were held at this temperature for a length of time sufficient to allow the composition gradients to be eliminated, the composition of the crystalline phase would reach α_2 and that of the liquid would reach L_2).

When the mixture is cooled to temperature T_3, the average composition of the solid and liquid phases will deviate from the equilibrium compositions as shown by the points α_3' and L_3'. Under equilibrium conditions T_3 is the temperature of final solidification, but as seen in the diagram, under nonequilibrium conditions of cooling a small quantity of melt remains until the sample temperature reaches T_4. Thus, when cooling is too rapid to permit attainment of equilibrium, the solidus and liquidus temperatures are decreased and the crystals are of nonuniform composition.

Bibliography and Supplementary Reading

I. A. Aksay and J. A. Pask, *J. Am. Ceram. Soc.,* **58** [11–12] 507 (1975).
G. H. Beall; High Temperature Oxides, Vol. 5-IV. Edited by A. M. Alper. Academic Press, New York, 1971.
C. G. Bergeron; Introduction to Glass Science. Edited by L. D. Pye, H. Stevens, and W. C. LaCourse, Plenum, New York, 1972.
R. F. Davis and J. A. Pask, *J. Am. Ceram. Soc.,* **55** [10] 525 (1972).
P. I. K. Onorato and D. R. Uhlmann, *J. Non-Cryst. Solids,* **22**, 367 (1976).
S. H. Risbud and J. A. Pask, *J. Mater. Sci.,* **13**, 2449 (1978).
T. P. Seward III; pp. 295–338 in Phase Diagrams: Materials Science and Technology, Vol. 6-I. Edited by A. M. Alper. Academic Press, New York, 1970.
D. R. Uhlmann, *J. Non-Cryst. Solids,* **7**, 337 (1972).
D. R. Uhlmann; pp. 80–124 in Advances in Ceramics, Vol. IV. Edited by J. H. Simmons, D. R. Uhlmann, and G. H. Beall. The American Ceramic Society, Columbus, Ohio, 1982.
J. White; Phase Diagrams: Materials Science and Technology, Vol. 6-II. Edited by A. M. Alper. Academic Press, New York, 1970.

Quaternary Systems   Chapter 8

8.1 System Representation

The four-component system may appear ominous at first encounter because of the seeming complexity of its representation. A comparison with ternary systems has been found to be helpful in developing the model used here.

Ternary systems require two dimensions to represent composition and a third dimension for temperature. The resulting figure is a prism such as shown in Fig. 8.1.

A plane projection of the liquidus surface (Fig. 8.2) is commonly used to represent the system. With the aid of the Alkemade theorem and other rules of construction, one can determine the solid and liquid phases that are present and their relative proportions for a given composition at a selected temperature.

In a quaternary system, three dimensions are required to represent composition and a fourth dimension is needed to represent temperature. Since we live in a three-dimensional world, this places us in the awkward position of being unable to construct a satisfactory figure to represent the quaternary system in a manner analogous to the prism of the ternary system. This dilemma is partially overcome by constructing a figure which is analogous to the *plane projection* of the liquidus surface of the ternary system. Such a figure is the tetrahedron shown in Fig. 8.3. Each corner represents the composition of one of the components of the system. The faces of this tetrahedron represent limiting ternary diagrams in a manner similar to that in which the faces of a solid ternary figure represent the limiting binary diagrams. The graphical representation employed in the ternary diagram (Fig. 8.2) can be extended to the solid model tetrahedron. The fields or areas of the ternary system become volumes in the tetrahedron, boundary lines on the ternary diagram extend into surfaces in the tetrahedron, and points on the limiting ternary faces become lines in the quaternary system.

An invariant point in the ternary diagram represents a point where three crystalline phases are in equilibrium with the melt. In a quaternary system this point is on a univariant line, on which these three crystalline phases are in equilibrium with melt whose composition varies along this line, and is deter-

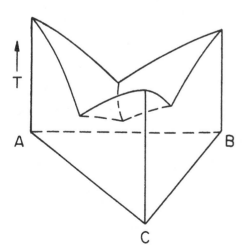

Fig. 8.1. Ternary system represented by a prism.

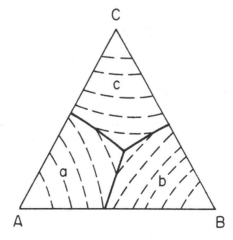

Fig. 8.2. Plane projection of the liquidus surface of a ternary system.

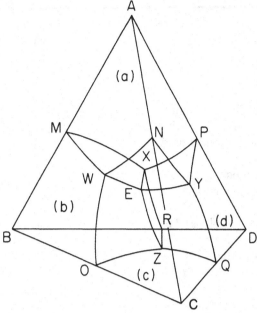

Fig. 8.3. Quaternary model of a system composed of four components A, B, C, and D: quaternary eutectic = E; ternary eutectics = W, X, Y, and Z; binary eutectics = M, N, O, P, Q, and R. Primary phase volumes: (a) bounded by sides $AMWN$-$ANYP$-$APXM$-$MWEX$-$WNYE$-$PYEX$; (b) bounded by sides $BMWO$-$BOZR$-$BRXM$-$OWEZ$-$RZEX$-$MWEX$; (c) bounded by sides $COWN$-$CQYN$-$COZQ$-$OZEW$-$QZEY$-$NWEY$; and (d) bounded by sides $DQYP$-$DQZR$-$DRXP$-$QZEY$-$PYEX$-$RZEX$.

mined with reference to the four apices of the tetrahedron. A boundary line on the ternary diagram expresses the coexistence of two phases in equilibrium with the melt. On the quaternary diagram, this line is on a surface (bivariant) on which these two crystalline phases are in equilibrium with the melt whose composition varies along the surface. The primary fields or areas of the ternary diagram extend into primary phase volumes in the quaternary diagram. Any point in these volumes represents one crystalline phase in equilibrium with a liquid whose composition is represented by this point. The surfaces of these primary volumes are curved boundary surfaces representing two crystalline phases in equilibrium with the melt. The intersection of three such surfaces formed when three phase volumes join each other forms a univariant line representing three crystalline phases in equilibrium with the melt. The junction of four primary phase volumes is a point where six curved surfaces meet and where four univariant lines intersect; this point is a quaternary invariant point at which four crystalline phases are in equilibrium with a melt whose composition is represented by this point. Figure 8.3 illustrates a quaternary model on which a quaternary eutectic, E, appears. In addition to this eutectic, there are four ternary eutectics, W, X, Y, and Z, and six binary eutectics, Q, R, O, P, N, and M, represented by this hypothetical system.

In the quaternary diagram, isotherms may be represented by curved surfaces extending from the isothermal lines of each of the limiting ternary faces of the tetrahedron into the interior of the figure (Fig. 8.4).

A join in a ternary system is represented by a line. In a quaternary system a join is a plane connecting three crystalline phases. Analogous to the

134

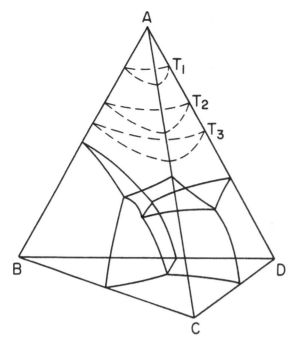

Fig. 8.4. Quaternary system with isotherms T_1, T_1, and T_3 added to primary phase volume of component A.

Alkemade line of a ternary system is the Alkemade plane (or compatibility join) of the quaternary system.

In ternary systems a compatibility triangle is made up of three Alkemade lines. In a quaternary system four Alkemade planes form a compatibility

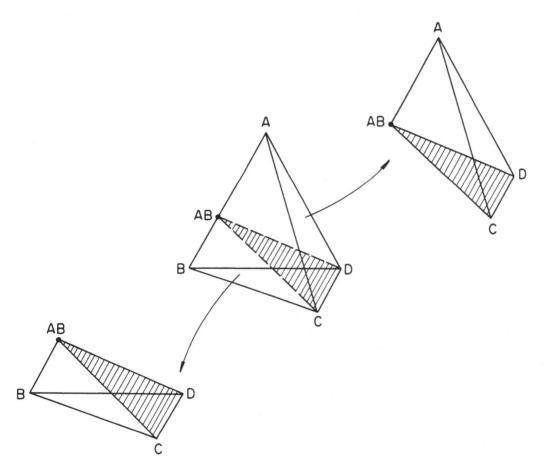

Fig. 8.5. Quaternary system with two compatibility tetraheda.

tetrahedron. In Fig. 8.5 the tetrahedra *AB-D-C-B* and *AB-A-D-C* are compatibility tetrahedra.

A quantitative isoplethal study in the system shown in Fig. 8.3 would require a three-dimensional model or a series of plane sections through the model in order to permit measurement of the lengths and positions of tie lines. The following example will serve to illustrate, in a qualitative way, a simple crystallization sequence.

Isoplethal Study

1. Final Crystals. In Fig. 8.6 composition *k* ($A = 61\%$, $B = 29\%$, $C = 6\%$,

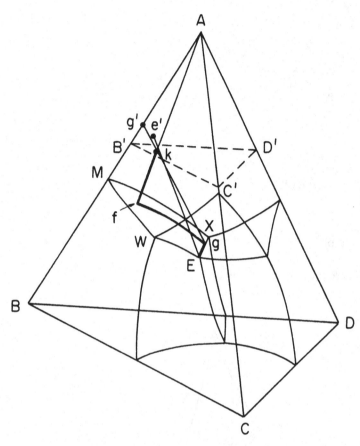

Fig. 8.6. Cooling path of melt of composition *k*: $A = 61$, $B = 29$, $C = 6$, and $D = 4$.

$D = 4\%$) is located within the compatibility tetrahedron *ABCD*. Because there are no intermediate compounds, there is only one compatibility tetrahedron in the system. When solidification is complete, the sample will be composed of crystals of A, B, C, and D. The proportions of A can be determined by constructing a plane through the point *k* and parallel to the base *BCD* of the compatibility tetrahedron. This plane is located at the $A = 61$ level. The intersection of this plane with the sides of the tetrahedron forms the triangle *B'C'D'*. The proportions of B, C, and D can be determined in the usual manner according to the position of the point *k* within the triangle. Note that *B'C'D'* represents a total of only 39% of the sample composition.

2. Cooling Path of Melt. A. The melt will cool to the point at which the crystals of A, B, C, and D are in equilibrium. This point is located at the quaternary eutectic *E*. Composition *k* is located within the primary phase volume of A; therefore, the first crystal to precipitate on cooling the melt will be A. As A precipitates from the melt, the concentration of A in the melt is decreased and the melt composition changes along a tie line drawn from A through point *k* and directly away from point A as shown in Fig. 8.6. This tie

136

line eventually intersects the boundary surface *MWEX*; the proportions of crystals of A and of melt are determined by applying the lever rule. The composition of the melt is determined by the position of the point *f* within the tetrahedron *ABCD*. Point *f* is located in the surface *MWEx*, which represents the composition of melts in equilibrium with crystals of A and B (surface *MWEx* is the common boundary between the primary phase volumes of A and B).

B. As the melt cools along the surface *MWEx*, one end of the tie line moves along the line from *A* to *g'* while the other moves along the line from *f* to *g*. The latter path describes the changing composition of the melt as crystals of A and B coprecipitate.

C. At point *g*, which is on the line *xE,* crystals of D begin to precipitate along with crystals of A and of B as the melt changes in composition from *g* to *E*. On further cooling at point *E*, the eutectic reaction occurs and the melt solidifies to crystals of A, B, C, and D in eutectic proportions.

3. Cooling Path of Solid. The compositional changes of the solid portion of the sample can be followed during cooling by tracing the intersections of successive tie lines with the appropriate Alkemade lines or planes. During the first stage of cooling, while crystals of A precipitate from the melt, the tie line terminates at point *A*. As the melt continues to cool, the end of the tie line moves from *A* to *g'* along the Alkemade line *AB* and crystals of B coprecipitate with crystals of A. On further cooling, crystals of D coprecipitate with crystals of A and B and the tie line intersects the Alkemade plane *ABD*. The end of the tie line moves from *g'* to point *e'*, which is located in the plane *ABD*. During the eutectic reaction the overall crystalline composition changes from that at point *e'* to that given by point *k*, which is the original composition point located within the tetrahedron.

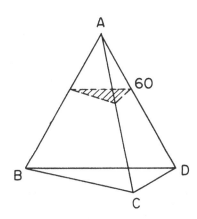

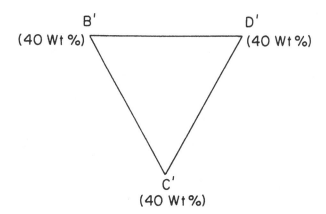

Fig. 8.7.   Method of investigating quaternary system.

137

8.2 Determination of Phase Relations

A method of investigating systematically the quaternary phase relations in ceramic systems generally involves a variation of three of the four components as the concentration of the fourth component is held constant. The selection of such compositional slices through the tetrahedron (Fig. 8.7) provides a simplified method of attack in the study of four-component systems. Often, phase studies of quaternary systems are made by dividing the system into smaller parts which represent smaller ternary or quaternary systems which are of particular interest to a segment of the industry or pertinent to a research investigation. Over a period of time, the system gradually becomes completely determined.

There are other methods which have been used to represent quaternary systems. A system described by Ricci involves the use of isothermal tetrahedra in which each tetrahedron represents all of the solid and liquid phases that are present at a given temperature.

Another method, described by Findlay, uses projections of lines and points onto the base of the composition tetrahedron to permit the construction of cooling paths and tie lines on a plane surface.

Phase relations in systems having more than four components can best be studied by a systematic variation of the composition variables. In a five-component (quinary) system, for example, two composition variables are held constant while the other three components are investigated as a ternary system. The two components are independently and progressively changed in a systematic manner in increments of 5 or 10%. Phase relations in systems composed of five or more components are difficult to represent graphically, but the important composition variables affecting the phase relations may be represented by ternary or quaternary diagrams with the effect of the other component or components schematically superimposed on these diagrams.

Relative Atomic Weights* **Appendix 1**

Symbol	Name	Atomic Number	Atomic Weight
Ac	Actinium	89	. . .
Ag	Silver	47	107.870
Al	Aluminum	13	26.9815
Am	Americium	95	. . .
Ar	Argon	18	39.948
As	Arsenic	33	74.9216
At	Astatine	85	. . .
Au	Gold	79	196.967
B	Boron	5	10.811
Ba	Barium	56	137.34
Be	Beryllium	4	9.0122
Bi	Bismuth	83	208.980
Bk	Berkelium	97	. . .
Br	Bromine	35	79.909
C	Carbon	6	12.01115
Ca	Calcium	20	40.08
Cd	Cadmium	48	112.40
Ce	Cerium	58	140.12
Cf	Californium	98	. . .
Cl	Chlorine	17	35.453
Cm	Curium	96	. . .
Co	Cobalt	27	58.9332
Cr	Chromium	24	51.996
Cs	Cesium	55	132.905
Cu	Copper	29	63.54
Dy	Dysprosium	66	162.50
Er	Erbium	68	167.26
Es	Einsteinium	99	. . .
Eu	Europium	63	151.96
F	Fluorine	9	18.9984
Fe	Iron	26	55.847
Fm	Fermium	100	. . .
Fr	Francium	87	. . .
Ga	Gallium	31	69.72
Gd	Gadolinium	64	157.25
Ge	Germanium	32	72.59
H	Hydrogen	1	1.00797
He	Helium	2	4.0026
Hf	Hafnium	72	178.49
Hg	Mercury	80	200.59
Ho	Holmium	67	164.930
I	Iodine	53	126.9044
In	Indium	49	114.82
Ir	Iridium	77	192.2
K	Potassium	19	39.102
Kr	Krypton	36	83.80
La	Lanthanum	57	138.91
Li	Lithium	3	6.939
Lu	Lutetium	71	174.97
Md	Mendelevium	101	. . .
Mg	Magnesium	12	24.312
Mn	Manganese	25	54.9381
Mo	Molybdenum	42	95.94

Symbol	Name	Atomic Number	Atomic Weight
N	Nitrogen	7	14.0067
Na	Sodium	11	22.9898
Nb	Niobium	41	92.906
Nd	Neodymium	60	144.24
Ne	Neon	10	20.183
Ni	Nickel	28	58.71
No	Nobelium	102	. . .
Np	Neptunium	93	. . .
O	Oxygen	8	15.9994 ±0.0001
Os	Osmium	76	190.2
P	Phosphorus	15	30.9738
Pa	Protactinium	91	. . .
Pb	Lead	82	207.19
Pd	Palladium	46	106.4
Pm	Promethium	61	. . .
Po	Polonium	84	. . .
Pr	Praseodymium	59	140.907
Pt	Platinum	78	195.09
Pu	Plutonium	94	. . .
Ra	Radium	88	. . .
Rb	Rubidium	37	85.47
Re	Rhenium	75	186.2
Rh	Rhodium	45	102.905
Rn	Radon	86	. . .
Ru	Ruthenium	44	101.07
S	Sulfur	16	32.064
Sb	Antimony	51	121.75
Sc	Scandium	21	44.956
Se	Selenium	34	78.96
Si	Silicon	14	28.086 ±0.001
Sm	Samarium	62	150.35
Sn	Tin	50	118.69
Sr	Strontium	38	87.62
Ta	Tantalum	73	180.948
Tb	Terbium	65	158.924
Tc	Technetium	43	. . .
Te	Tellurium	52	127.60
Th	Thorium	90	232.038
Ti	Titanium	22	47.90
Tl	Thallium	81	204.37
Tm	Thulium	69	168.934
U	Uranium	92	238.03
V	Vanadium	23	50.942
W	Tungsten	74	183.85
Xe	Xenon	54	131.30
Y	Yttrium	39	88.905
Yb	Ytterbium	70	173.04
Zn	Zinc	30	65.37
Zr	Zirconium	40	91.22

*Based on $^{12}C = 12$.

139

Molecular Weights of Oxides

Oxide	Molecular Wt.	Oxide	Molecular Wt.	Oxide	Molecular Wt.
Ag_2O	231.739	MgO	40.311	Sb_2O_3	291.50
Al_2O_3	101.9612	MnO	70.9375	Sb_2O_5	323.50
Au_2O	409.933	MnO_2	86.9369	Sc_2O_3	137.910
Au_2O_3	441.932	Mn_2O_3	157.8744	SeO_2	110.96
As_2O_3	197.8414	Mn_3O_4	228.8119	SeO_3	126.96
As_2O_5	229.8402	MoO_2	127.94	SiO_2	60.085
		MoO_3	143.94	Sm_2O_3	348.70
B_2O_3	69.620	Mo_2O_3	239.88	SnO	134.69
BaO	153.34			SnO_2	150.69
BeO	25.0116	Na_2O	61.9790	SO_2	64.063
Bi_2O_3	465.958	Nb_2O_5	265.809	SO_3	80.062
		Nd_2O_3	336.48	SrO	103.62
CaO	56.08	NiO	74.71		
CdO	128.40	Ni_2O_3	165.42		
CeO_2	172.12	Ni_3O_4	240.13	Ta_2O_5	441.893
Ce_2O_3	328.24	NO	30.0061	Tb_2O_3	365.846
CO	28.0106	NO_2	46.0055	Tb_4O_7	747.692
CO_2	44.0100	N_2O	44.0128	TeO_2	159.60
CoO	74.9326			TeO_3	175.60
Co_2O_3	165.8646	OsO	206.2	ThO_2	264.037
CrO_2	83.995	OsO_2	222.2	TiO	63.90
CrO_3	99.994	OsO_4	254.2	TiO_2	79.90
Cr_2O_3	151.990	Os_2O_3	428.4	Ti_2O_3	143.80
Cs_2O	281.809			Ti_3O_5	223.70
CuO	79.54	PbO	223.19	Tl_2O	424.74
Cu_2O	143.08	PbO_2	239.19	Tl_2O_3	456.74
		Pb_2O	430.38	Tm_2O_3	385.866
Dy_2O_3	373.00	Pb_2O_3	462.38		
		Pb_3O_4	685.57		
Er_2O_3	382.52	PdO	122.4	UO	254.03
Eu_2O_3	351.92	PdO_2	138.4	UO_2	270.03
		P_2O_5	141.9446	UO_3	286.03
FeO	71.846	PrO_2	172.906	U_2O_5	556.06
Fe_2O_3	159.692	Pr_2O_3	329.812	U_3O_7	826.09
Fe_3O_4	231.539	Pr_6O_{11}	1021.435	U_3O_8	842.09
		PtO	211.09		
Ga_2O_3	187.44	PtO_2	227.09	VO	66.941
Gd_2O_3	362.50			V_2O_3	149.882
GeO_2	104.59	Rb_2O	186.94	V_2O_4	165.882
		ReO_2	218.2	V_2O_5	181.881
H_2O	18.0153	ReO_3	234.2		
H_2O_2	34.0147	Re_2O_3	420.4	WO_2	215.85
HfO_2	210.49	Re_2O_7	484.4	WO_3	231.85
HgO	216.59	RhO	118.904	W_4O_{11}	911.39
Hg_2O	417.18	RhO_2	134.904		
Ho_2O_3	377.858	Rh_2O_3	253.808	Yb_2O_3	394.08
		RuO	117.07	Y_2O_3	225.808
In_2O_3	277.64	RuO_2	133.07		
I_2O_5	333.8058	RuO_3	149.07	ZnO	81.37
IrO_2	224.2	RuO_4	165.07	ZrO_2	123.22
Ir_2O_3	432.4	Ru_2O_3	250.14		
K_2O	94.203				
La_2O_3	325.82				
Li_2O	29.877				
Lu_2O_3	397.94				

Oxide	Investigator	Environment	Melting Point Original °C	Int. 1948 °C
Ag₂O	F. C. Kracek	......	191D	...
	Shun-ichiro Iijima	......	230D	230D
Al₂O₃	E. Tiede & E. Birnbrauer	Vacuum	1890	...
	R. F. Geller & P. J. Yavorsky	(Air)	2000–2030	1994–2024
	O. Ruff & G. Lauschke	Air at 7.5 mm Hg	2005	2005
	" " "	Air at 7.7 mm Hg	2008	2008
	O. Weigel & F. Kaysser	N₂ at 1 atm	2007	2009
	" " "	Air	2010	2012
	" " "	"	2005–2010	2007–2012
	" " "	Reducing	2001	2003
	S. D. Mark, Jr.	Neutral	2020	2020
	J. J. Diamond & S. J. Schneider	Air	2025	2025
	R. F. Geller & E. N. Bunting	Air	2035	2029
	W. A. Lambertson & F. H. Gunzel, Jr.	He	2034	2034
	O. Ruff	......	2010	2035
	E. N. Bunting	(Air)	2045	2038
	R. H. McNally, R. I. Peters, & P. H. Ribbe	Air, Ar, N₂	2043	2043
	O. Ruff & O. Goecke	N₂	2020	2044
	S. M. Lang, F. P. Knudsen, C. L. Fillmore & R. S. Roth	Ar	2049	2049
	H. v. Wartenberg, H. Lindy, & R. Jung	Air	2055	2049
	C. W. Kanolt	2 mm Hg; H₂	2050	2072
As₂O₃	A. Smits and E. Beljaar	66.1 mm Hg	312.3	312.3
	H. V. Welsch & L. H. Duschak	......	313	...
	E. R. Rushton & F. Daniels	......	315	...
B₂O₃	F. C. Kracek, G. W. Morey, & H. E. Merwin	(Air)	450	450
	L. McCullock	......	460–470	...
BaO	E. E. Schumacher	H₂ at 0.2 atm	1923	1918
BeO	E. Tiede & E. Birnbrauer	Vacuum	2400	...
	O. Ruff & G. Lauschke	15 mm Hg	2410	2410
	S. M. Lang, F. P. Knudsen, C. L. Fillmore, & R. S. Roth	Ar	2452	2452
	H. v. Wartenberg, H. J. Reusch & E. Saran	Oxidizing	2520	2508
	H. v. Wartenberg & H. Werth	Oxidizing	2570	2557
	Ya I. Ol'shanskiĭ	N₂	2570	2570
	O. Ruff	N₂ at 4–10 mm Hg	2525	2573
Bi₂O₃	L. Belladen	......	817	...
	W. Guertler	......	820	...
	G. Gattow & H. Schröder	(Air)	824	824
	E. M. Levin & C. L. McDaniel	Air	825	825

Oxide	Investigator	Environment	Melting Point Original °C	Int. 1948 °C
CaO	E. E. Schumacher	H₂ at 0.2 atm	2576	2565
	C. W. Kanolt	H₂	2572	2614
	Ya I. Ol'shanskiĭ	N₂	2620	2620
	R. C. Doman, J. B. Barr, N. R. McNally, & A. M. Alper	......	2630	2630
CdO	R. S. Roth	(Air)	>1500	>1500
CeO₂	O. Ruff	......	1950	1973
	H. v. Wartenberg & W. Gurr	Air	>2600	>2600
	F. Trombé	......	2800	...
CoO	H. v. Wartenberg, H. J. Reusch, & E. Saran	Oxidizing	1800	1795
	H. v. Wartenberg & E. Prophet	Air	1810	1805
Cr₂O₃	O. Ruff	N₂ at 30 mm Hg	1830–2080	1849–2107
		N₂ at 1 atm	1960	1983
	C. W. Kanolt	Vacuum	1990	2011
	W. T. Wilde & W. J. Rees	Air	2060	2053
	E. N. Bunting	(Air)	2275	2266
	H. v. Wartenberg & H. J. Reusch	Air	2275	2266
	R. H. McNally, R. I. Peters, & P. H. Ribbe	N₂	2315	2315
	H. v. Wartenberg & K. Eckhardt	Air	2330	2330
		Air	2435	2424
Cs₂O	M. E. Rengade	N₂	490	...
Cu₂O	R. Ruer & M. Nakamoto	N₂	1222	1222
	H. v. Wartenberg, H. J. Reusch, & E. Saran	Oxidizing	1230	1229
	H. S. Roberts & F. H. Smyth	0.6 mm Hg	1235	1236
Dy₂O₃	L. G. Wisnyi & S. Pijanowski	Either He, H₂ or vacuum	2340	2340
Eu₂O₃	L. G. Wisnyi & S. Pijanowski	Either He, H₂ or vacuum	2050	2050
	S. J. Schneider	Air	2240	2240
FeO	J. Chipman & S. Marshall	Slightly oxidizing	1369	1368
	R. Hay, D. D. Howat & J. White	N₂ at 1 atm	1370	1368
	L. S. Darken & R. W. Gurry	N₂ at 1 atm	1371	1369
	N. L. Bowen & J. F. Schairer	N₂-slightly oxidizing	1380	1382
Fe₃O₄	V. L. Moruzzi & M. W. Shafer	Air	1591	1591
	L. S. Darken & R. W. Gurry	O₂ at 1 atm	1583	1580
		Air	1594	1591
		O₂ at 0.0575 atm	1597	1594
	J. W. Greig, E. Posnjak, H. E. Merwin, & R. B. Sosman	Small O₂ pressure	1591	1594
	H. v. Wartenberg & K. Eckhardt	Air	1650	1647
Ga₂O₃	V. G. Hill, R. Roy, & E. F. Osborn	......	1725	1725
	H. v. Wartenberg & H. J. Reusch	Air	1740	1736
	S. J. Schneider & J. L. Waring	Air	1795	1795
Gd₂O₃	L. G. Wisnyi & S. Pijanowski	Either He, H₂ or vacuum	2330	2330
	C. E. Curtis & J. R. Johnson	Air	2350	2350
GeO₂	R. Schwarz, P. W. Schenk, & H. Giese	Air	1115	1115
	A. W. Laubengayer & B. S. Morton	(Air)	1116	1116
HfO₂	P. Clausing	H₂	2774	2758
	S. D. Mark, Jr.	Neutral	2770	2770
	F. Henning	N₂ or H₂	2812	...
	C. E. Curtis, L. M. Doney, & J. R. Johnson	......	2900	2900
In₂O₃	S. J. Schneider	Air	1910	1910

Oxide	Investigator	Environment	Melting Point Original °C	Int. 1948 °C
IrO$_2$	E. H. P. Cordfunke & G. Meyer	O$_2$ at 1 atm	1100D	1100D
La$_2$O$_3$	O. Ruff	......	1840	1859
	W. A. Lambertson & F. H. Gunzel, Jr.	H$_2$	2210	2210
	H. v. Wartenberg & H. J. Reusch	Air	2315	2307
Li$_2$O	Handbook of Chemistry and Physics	......	>1700	...
MgO	O. Ruff	N$_2$ at 10–30 mm Hg	2120–2550	2150–2599
		N$_2$ at 1 atm	2250–2500	2285–2546
	K. K. Kelley	......	2642	...
	R. H. McNally, R. I. Peters, & P. H. Ribbe	N$_2$	2825	2825
	C. W. Kanolt	CO and N$_2$ at 1 atm	2800	2852
MnO	E. Tiede & E. Birnbrauer	Vacuum	1650	...
	J. White, D. D. Howat, and R. Hay	......	1785	1781
Mn$_3$O$_4$	H. v. Wartenberg & E. Prophet	Air	1560	1557
	H. J. Van Hook & M. L. Keith	Air	1562	1564
	T. Ranganathan, B. E. MacKean, & A. Muan	Air	1567	1567
	H. v. Wartenberg, H. J. Reusch, & E. Saran	Air	1590	1587
	H. v. Wartenberg & W. Gurr	Air	1705	1701
MoO$_3$	T. Carnelley	......	759	...
	E. Groschuff	Air	791	...
	F. M. v. Jaeger & H. C. Germs	Oxidizing	795	...
	F. Hoermann	......	795	795
	G. D. Rieck	Air	795	795
	L. A. Cosgrove & P. E. Snyder	N$_2$ at 1 atm	795.4	795.4
Nb$_2$O$_5$	G. Brauer	O$_2$	1460	1458
	M. W. Shafer & R. Roy	......	1465	1465
	M. Ibrahim, N. F. Bright & J. F. Rowland	Air	1479	1479
	R. S. Roth & J. L. Waring	Air	1485	1485
	A. Reisman & F. Holtzberg	O$_2$ + Air	1486	1486
	R. S. Roth & J. L. Waring	Air	1487	1487
	F. Holtzberg, A. Reisman, M. Berry, & M. Berkenblit	(Air)	1491	1491
	J. J. Diamond & S. J. Schneider	Air	1496	1496
	R. S. Roth & L. W. Coughanour	Air	1500	1500
	R. L. Orr	......	1512	1512
	O. Ruff	......	1520	1530
Nd$_2$O$_3$	W. A. Lambertson & F. H. Gunzel	He	2272	2272
NiO	P. D. Merica & R. G. Waltenberg	Vacuum	1552	...
		Air	1660	...
	H. v. Wartenberg & E. Prophet	Air	1990	1984
OsO$_4$	H. v. Wartenberg	≈11 mm Hg	40.1	...
	Handbook of Chemistry & Physics	......	41	...
	K. K. Kelley	......	56	...
P$_2$O$_5$	J. M. A. Hoeflake & M. F. C. Scheffer	4600 mm Hg	569	569
PbO	L. Belladen	......	870	...
	S. Hilpert & P. Weiller	Air	876	...
	F. M. v. Jaeger & H. C. Germs	Oxidizing	877	...
	R. Schenck & W. Rassbach	Air	879	...
	R. F. Geller, A. S. Creamer, & E. N. Bunting	Air	886	886
	K. A. Krakau	Air	886	886
	V. A. Kroll	Air	888	...
	H. C. Cooper, L. I. Shaw, & N. E. Loomis	(Air)	888	...
PtO$_2$	Handbook of Chemistry & Physics	......	450	...
Re$_2$O$_7$	K. K. Kelley	......	296	...
Sb$_2$O$_3$	W. B. Hincke	8.5 mm Hg	655	655

Oxide	Investigator	Environment	Melting Point Original °C	Int. 1948 °C
Sc_2O_3	S. J. Schneider & J. L. Waring	Air	>2405	>2405
SeO_2	Handbook of Chemistry & Physics	......	340–350	...
SiO_2	R. Weitzel	Air	1696	1691
	K. Endell & R. Rieke	N_2	1685	1692
	J. White, D. D. Howat, & R. Hay	......	1705	1701
	J. B. Ferguson & H. E. Merwin	Air	1710	1720
	J. W. Greig	Air	1713	1723
	N. Zhirnova	Air	1715	1728
	O. Ruff & G. Lauschke	17 mm Hg	1850	1850
Sm_2O_3	L. G. Wisnyi & S. Pijanowski	Either He, H_2 or vacuum	2300	2300
	C. E. Curtis & J. R. Johnson	Air	2350	2350
SnO_2	O. Ruff	......	1385	1391
		......	1625	1637
	V. J. Barczak & R. H. Insley	......	1630	1630
SrO	E. E. Schumacher	H_2 at 0.2 atm	2430	2420
Ta_2O_5	A. Reisman, F. Holtzberg, M. Berkenblit & M. Berry	(Air)	1872	1872
	O. Ruff	N_2 at reduced P	1875	1895
TeO_2	F. C. Kracek	......	732.6	...
ThO_2	E. Tiede & E. Birnbrauer	Vacuum	2000	...
	O. Ruff	N_2 at reduced P	2425	2468
			2440	2483
			2470	2515
	F. Trombé	......	3000	...
	O. Ruff, F. Ebert, & H. Woitinek	......	3050	3030
	W. A. Lambertson & F. H. Gunzel, Jr.	He	3220	3220
TiO_2	W. O. Statton	Vacuum	1720	1716
	H. v. Wartenberg & E. Prophet	Air	1825	1820
	H. Sigurdson & S. S. Cole	Oxidizing	1825	1825
	D. E. Rase & R. Roy	(Air)	1830	...
	L. W. Coughanour & V. A. DeProsse	Air	1839	1839
	P. D. S. St. Pierre	Air	1840	1840
	J. J. Diamond & S. J. Schneider	Air	1840	1840
	H. v. Wartenberg & W. Gurr	Air	1850	1845
	S. M. Lang, C. L. Fillmore & L. H. Maxwell	Air	1845	1845
	H. v. Wartenberg & K. Eckhardt	Air	1855	1850
	G. Brauer & W. Littke	O_2 at 300 torr + Ar at 460 torr	1840	1840
		O_2 at 500 torr + Ar at 260 torr	1860	1860
		O_2 at 600 torr + Ar at 160 torr	1870	1870
		O_2 at 760 torr	1870	1870
		O_2 at 1140 torr	1870	1870
Tl_2O_3	A. B. F. Duncan	O_2 at 1 atm	717	717
UO_2	O. Ruff & O. Goecke	N_2	2176	2208
	L. G. Wisnyi & S. Pijanowski	Either He, H or vacuum	2760	2760
	T. C. Ehlert & J. L. Margrave	Vacuum	2860	2860
	W. A. Lambertson & F. H. Gunzel, Jr.	He	2878	2878
V_2O_5	F. C. Kracek	......	656	...
	T. Carnelley	......	658	...
	O. A. Cook	......	670	670
	V. V. Illarionov, R. P. Ozeron, & E. V. Kil'disheva	......	672	672
	F. Holtzberg, A. Reisman, M. Berry, & M. Berkenblit	Air + O_2	674	674
	C. McDaniel	Air	675	675
	A. Burdese	CO_2	685	685
	Handbook of Chemistry & Physics	......	690	...

Oxide	Investigator	Environment	Melting Point Original °C	Int 1948 °C
WO$_3$	F. M. v. Jaeger & H. C. Germs	Oxidizing	1473	...
	F. Hoermann	......	1473	1471
Y$_2$O$_3$	O. Ruff & G. Lauschke	Air at 21.5 mm Hg	2410	2410
	O. Ruff	N$_2$ at 15 mm Hg	2415	2458
ZnO	E. N. Bunting	Air	1975	1969
ZrO$_2$	E. Tiede & E. Birnbrauer	Vacuum	2430	...
	O. Ruff & G. Lauschke	H$_2$ at 1 atm	2519	2519
		Air at 8.22 mm Hg	2563	2563
	O. Ruff	N$_2$	2585	2636
	P. Clausing	H$_2$	2677	2663
	E. Podszus	Air	2677	...
		"	2727	...
	F. Henning	H$_2$ and N$_2$	2687	...
	S. D. Mark, Jr.	Neutral	2690	2690
	F. Trombé	......	2700	...
	W. A. Lambertson & F. H. Gunzel, Jr.	He	2710	2710
	N. Zhirnova	Air	2715	2765
	C. E. Curtis, L. M. Doney, & J. R. Johnson	......	2850	2850

Considerable activity in research, development, and processing of modern ceramics is based on nonoxide ceramic materials. Conventional oxide ceramics have been used for many decades, and much phase equilibria information thus exists for these systems. Since nonoxide ceramics are of relatively recent vintage, phase data on these systems, when available, are scattered in the literature. In keeping with the philosophy that any material that is not organic or metallic is a prospective technological ceramic, we have included a sprinkling of phase diagrams of some nonoxide materials in this Appendix. The selection of specific systems is somewhat random, although admittedly biased by the current emphasis on nitride, fluoride, carbide, and the Group III–V and related compound semiconducting materials.

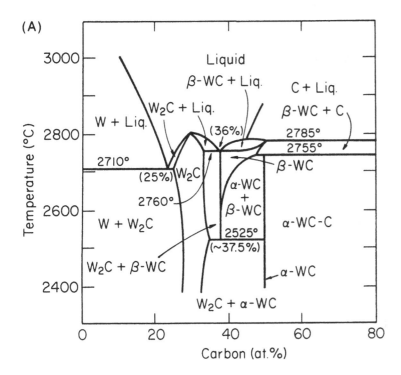

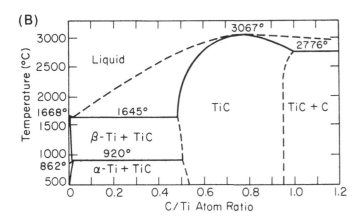

Fig. A.1. Phase diagrams of the system (A) W-C (from Toth (1971)) and (B) the system Ti-C (from Storms (1967)). (Reprinted by permission.)

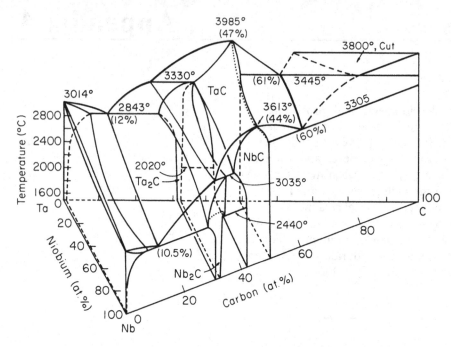

Fig. A.2. Phase diagram of the ternary system Nb-Ta-C (from Rudy (1969)).

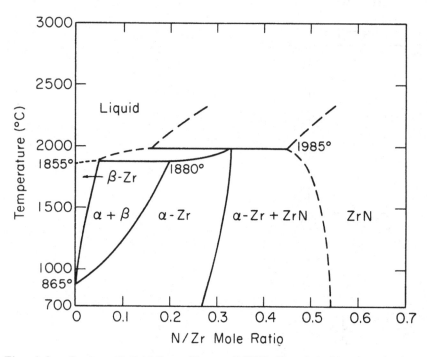

Fig. A.3. System Zr-ZrN (from Storms (1967)). (Reprinted by permission.)

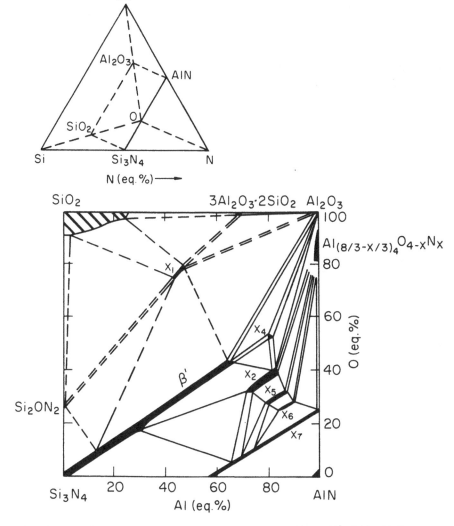

Fig. A.4. Phase equilibria in the subsection Si_3N_4-AlN-SiO_2-Al_2O_3 at an isotherm of 1760°C; dashed lines show tentative equilibria at a temperature slightly below 1760°C (from Gauckler et al. (1975)).

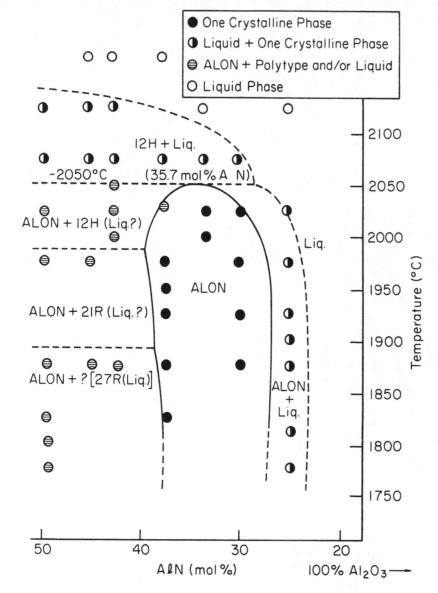

Fig. A.5. High-temperature phase equilibria (in 1 atm of flowing N_2) in the system Al_2O_3-AlN (from McCauley and Corbin (1979)).

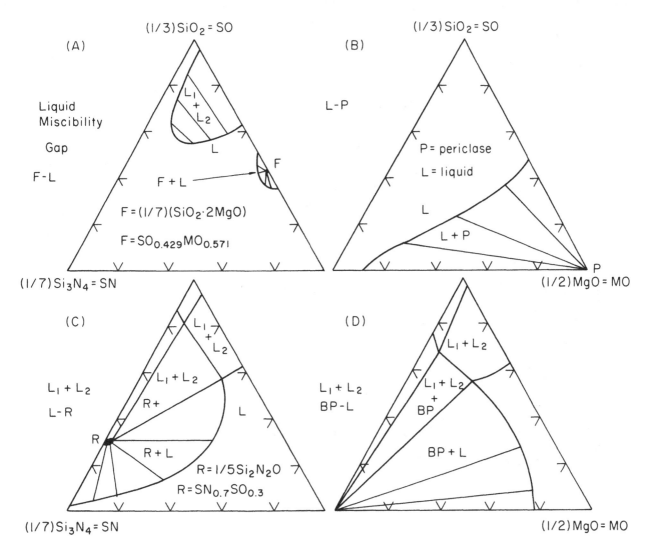

Fig. A.6. Calculated equilibria in the oxynitride system SiO_2-MgO-Si_3N_4 at 2100 K (after Kaufman et al. (1981)). (Reprinted by permission.)

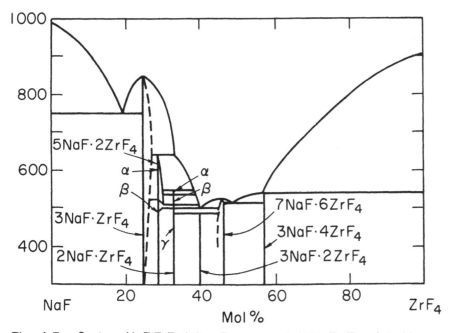

Fig. A.7. System NaF-ZrF$_4$ (after Barton et al. (1958)). (Reprinted by permission.)

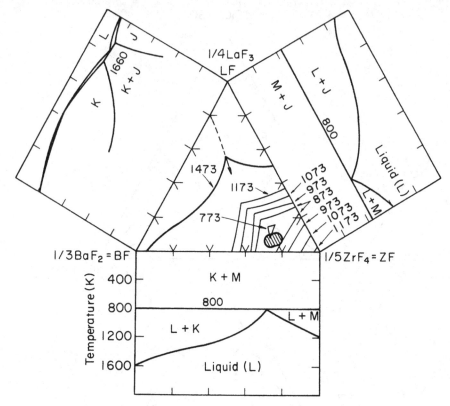

Fig. A.8. Liquidus projections in the system ZrF₄-LaF₃-BaF₂ as calculated by Kaufman et al. (1983). The hatched area shows region of glass formation as established experimentally by Lecoq and Poulain (1980)). (Reprinted by permission.)

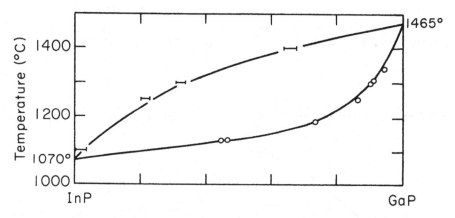

Fig. A.9. Quasi-binary system GaP-InP (from Panish and Ilegems (1972) and Foster and Scardefield (1970)). (Reprinted by permission.)

154

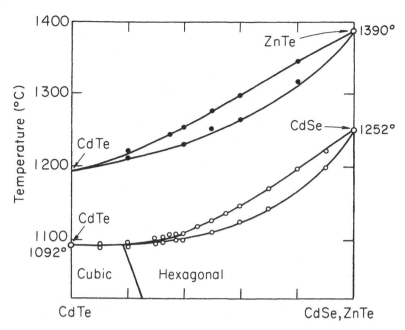

Fig. A.10. Quasi-binary systems CdTe-CdSe and CdTe-ZnTe (from Steininger et al. (1970) and Strauss and Steininger (1970)). (Reprinted by permission.)

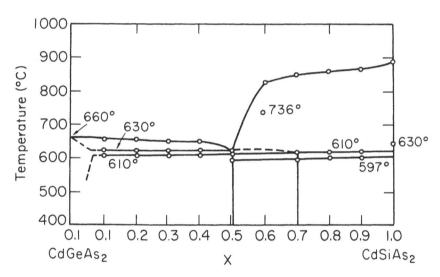

Fig. A.11. Melting relations diagram in the pseudobinary semiconducting system CdGeAs₂-CdSiAs₂ (from Bhattacherjee and Risbud (1982)). (Reprinted by permission.)

Bibliography

C. J. Barton, W. R. Grimes, H. Insley, R. E. Moore, and R. E. Thoma, *J. Phys. Chem.*, **62**, 666 (1958).

A. P. Bhattacherjee and S. H. Risbud, *Philos. Mag., [Part A]*, **46**, 995 (1982).

L. M. Foster and J. E. Scardefield, *J. Electrochem. Soc.*, **117**, 534 (1970).

L. J. Gauckler, H. L. Lukas, and G. Petzow, *J. Am. Ceram. Soc.*, **58** [7–8] 346 (1975).

L. Kaufman, J. Agren, J. Nell, and F. Hayes, *CALPHAD: Comput. Coupling Phase Diagrams Thermochem.*, **7**, 71 (1983).

L. Kaufman, F. Hayes, and D. Birnie, *CALPHAD: Comput. Coupling Phase Diagrams Thermochem.*, **5**, 163 (1981).

A. Lecoq and M. Poulain, *J. Non-Cryst. Solids*, **41**, 209 (1980).

J. W. McCauley and N. D. Corbin, *J. Am. Ceram. Soc.*, **62** [9–10] 476 (1979).

M. B. Panish and M. Ilegems, *Prog. Solid State Chem.*, **7**, 39 (1972).

E. Rudy, "Compendium of Phase Diagram Data," AFML-TR-65-2, Part V (1969).

J. Steininger, A. J. Strauss, and R. F. Bregick, *J. Electrochem. Soc.*, **117**, 1305 (1970).

E. K. Storms; The Refractory Carbides, Vol. 2. Academic Press, New York, 1967.

A. J. Strauss and J. Steininger, *J. Electrochem. Soc.*, **117**, 1420 (1970).

L. E. Toth: Transition Metal Carbides and Nitrides, Vol. 7. Academic Press, New York, 1971.

Index